Studies in Systems, Decision and Control

Volume 122

Series editor

Janusz Kacprzyk, Polish Academy of Sciences, Warsaw, Poland
e-mail: kacprzyk@ibspan.waw.pl

About this Series

The series "Studies in Systems, Decision and Control" (SSDC) covers both new developments and advances, as well as the state of the art, in the various areas of broadly perceived systems, decision making and control- quickly, up to date and with a high quality. The intent is to cover the theory, applications, and perspectives on the state of the art and future developments relevant to systems, decision making, control, complex processes and related areas, as embedded in the fields of engineering, computer science, physics, economics, social and life sciences, as well as the paradigms and methodologies behind them. The series contains monographs, textbooks, lecture notes and edited volumes in systems, decision making and control spanning the areas of Cyber-Physical Systems, Autonomous Systems, Sensor Networks, Control Systems, Energy Systems, Automotive Systems, Biological Systems, Vehicular Networking and Connected Vehicles, Aerospace Systems, Automation, Manufacturing, Smart Grids, Nonlinear Systems, Power Systems, Robotics, Social Systems, Economic Systems and other. Of particular value to both the contributors and the readership are the short publication timeframe and the world-wide distribution and exposure which enable both a wide and rapid dissemination of research output.

More information about this series at http://www.springer.com/series/13304

Sayed Hadi Sadeghi

E-Learning Practice in Higher Education: A Mixed-Method Comparative Analysis

Springer

Sayed Hadi Sadeghi
Quakers Hill, NSW
Australia

ISSN 2198-4182 ISSN 2198-4190 (electronic)
Studies in Systems, Decision and Control
ISBN 978-3-319-65938-1 ISBN 978-3-319-65939-8 (eBook)
https://doi.org/10.1007/978-3-319-65939-8

Library of Congress Control Number: 2017949494

Printed on acid-free paper

This Springer imprint is published by Springer Nature
The registered company is Springer International Publishing AG
The registered company address is: Gewerbestrasse 11, 6330 Cham, Switzerland

To my Dad Martyr Sayed Ali Sadeghi who taught me love and faith,

To my kind Mom who sacrificed her life to my success and my progress

and

To my Dear wife Zeynab for her loving support.

Acknowledgements

And [that] Allah may aid you with a mighty victory. (Quran, 3/48)

First and foremost, I am deeply grateful to my kind supervisor Associate Professor Nigel Bagnall. His remarkable insights and rich supervisory experience have benefited me in all steps of my Ph.D. Nigel also showed his faith in me with his unfailing encouragement. My most sincere thanks go to Nigel. Thank you also to my co-supervisor, Prof. Michael J. Jacobson, for his guidance and effective feedbacks.

Thank you to Prof. Kristin Anderson, Prof. Michelle Lincoln, and Dr. Bob Rubinyi, who provided support to allow me to collect my data from USA and Australia.

Thank you to Associate Professor Hadi Bahrami Ehsan, Dr. Mohammad Khansari, Dr. Davoud Masoumi, Dr. Lina Markauskaite, and Dr. Mohsen Paknejad for their advice in conducting this research program, and special thanks should also go to Ms. Tigger Wise for her professional editing.

I wish to thank the staff at the Faculty of Education and Social Work, specially Ms. Venice Jureidini-Briozzo and Ms. Suin Jung for supporting my study.

Finally, my thanks to my sister Sayed Fatemeh Sadeghi and God's gift of my little angel Sayed Mir Mohammad Sadeghi, who gives me hope, joy, and sweetness in my life.

Contents

Abstract

The overall aim of this study was to further comparative understanding of e-practice in Australian and American universities. The study used one Faculty in an Australian university and one Faculty in an American university as examples. The theoretical focus was on the cultural context as well as on practice in the e-learning area. The variables of instructivism and constructivism were explored for establishing the differing cultural contexts of the two countries. The pedagogical, performance appraisal, instructional design, technological, administrative, and support service were investigated to establish e-practice differences between the two countries. Studies 1 to 3 used both a qualitative and a quantitative methodology in order to ascertain the current status of e-learning. Participants were students, lecturers, and administrative staff of one Faculty in an Australian university and one Faculty in an American university engaged with e-learning programs. Study 1 investigated the dominant cultural dimensions of the two universities. The results of this study showed that the dominant e-learning approach of one Faculty in an American university was toward constructivism while the dominant e-learning approach of one Faculty in an Australian university was toward instructivism. In Study 2, the current status of e-learning practice was investigated in the two universities using a quantitative methodological approach.

The results indicated that the level of e-practice in all aspects of e-learning was above average in both universities. Participants of the American university rated their system consistently higher in most aspects of e-practice than the Australian university participants. In Study 3, the current issues of e-learning practice in four aspects, namely pedagogy, culture, technology, and e-practice, that need to be improved were investigated by applying a qualitative method. The results of interviews identified pedagogical challenges in approaches to learning, effective learning practice, assessment method, and learning content as areas that need attention. Cultural sensitivity, effective cultural practice, and key technological challenges as well as issues such as faculty policies, quality, learning management system, and online support were revealed as areas that could improve the e-learning systems in both universities. Although both America and Australia have shown progress in the field of e-practice, it is apparent that the quality and quantity of e-practice factors in an Australian university needs to be sped up. This is despite the fact that the context of e-learning in an Australian university studied has been improved by Asian cultural contact. From this perspective, applying the pattern and technology that has been used in the American university could help to guide an Australian university e-learning system practice in the future.

About the Author

Dr. Sayed Hadi Sadeghi obtained his B.Sc. and M.Sc. degrees in educational administration and planning from the University of Tehran in 2004 and 2007, respectively. In February 2016, he submitted his Ph.D. to the Faculty of Education and Social Work at the University of Sydney in the field of e-learning practice in higher education. He developed numerous project management skills over the course of his Ph.D. training. On the technical side, he became expert at mixed-method research design, applying SPSS and Nvivo for analyzing the data. While completing his Ph.D., he worked in an international educational assembly as Lecturer and Senior Researcher. His responsibilities in this role were identification and evaluation of major cultural events based on pedagogical context in Sydney online and off-line communities. Recently, he was appointed to the Editorial Board of the International Journal of Academic Studies. His responsibilities in this role are assessing the quality and validity of authors' manuscripts on computer-supported learning and e-learning practices. Since 2012, as member of the Oceania Comparative and International Education Society, he has attempted to expand the understanding of cultural–pedagogical paradigms in different cultures to improve the quality of e-practices, which fit well with his recent comparative study of e-learning between The University of Minnesota and The University of Sydney on practices and policies. For this, he received "A" grade and reference support letter from Pro Vice-chancellor Portfolio and Adjunct Professor at the School of

Education in UNSW, who was his examiner. He is working now as Project Director of the University of Tehran's human resource development (HRD) project, which is a new cognitive behavioral project to advance future management and organizational learning practice.

Chapter 1
Introduction and Overview

This study was motivated by an interest in comparing the practice of e-learning in higher education between one Faculty in an Australian university and one Faculty in an American university. To reach this aim, three studies were conducted in two high ranking universities that have provided e-learning courses in various fields. Accordingly, the cultural dimensions of their educational paradigms, their educational e-practice and some current issues common to both institutions have been compared. In this section of the introduction, a brief background of each component of studies is described, then aims, significance and questions are presented.

The main aim of e-learning in developed countries like Australia and the United States of America is to promote sustained quality improvement, to cultivate an operative knowledge economy and to increase the lifetime of pedagogical practice (Gulati 2008). Selim (2007) believes that: "The efficient and effective use of IT in delivering the e-learning based components of a course is of critical importance to the success and student acceptance of e-learning" (p. 399). Online courses provide a borderless market for universities and colleges without adding pressure to on-campus infrastructure, however, the capability, reliability and richness of the university IT infrastructure to deliver the courses as smoothly as possible are the key to the success of e-learning (Parsazadeh et al. 2013; Selim 2007).

A recent IBIS World report on online program revenue over the last 5 years highlights that the revenue from online programs grew dramatically by at least 80% from 2008 to 2012 (Barber 2013). According to the American National Center for Education Statistics, there has been substantial pedagogical restructure based on increasing online education opportunities. The number of US students enrolled in at least one e-learning program increased from 1 million to 12 million between 2002 and 2006. From this perspective, approximately 33% of students enrolled in higher education since 2007 have been interested to take at least one course online (Allen and Seaman 2011). According to the evidence that Allen and Seaman (2011) explained:

"After remaining steady for several years, the proportion of chief academic officers saying that online education is critical to their long-term strategy took an

S.H. Sadeghi, *E-Learning Practice in Higher Education: A Mixed-Method Comparative Analysis*, Studies in Systems, Decision and Control 122,
https://doi.org/10.1007/978-3-319-65939-8_1

upward turn in both 2010 and 2011. Sixty-five percent of all reporting institutions said that online learning was a critical part of their long-term strategy, a small increase from 63% in 2010. The year-to-year change was greatest among the for-profit institutions, whose agreement with this increased from 51% in 2009 to 69% in 2011. For-profit institutions are the most likely to have included online learning as a part of their strategic plan" (p. 4).

In 2011, Sloan Consortium (Sloan-C) issued a report on e-learning courses in American universities. This report revealed that the number of learners taking at least one e-learning program has now surpassed 6 million. Also nearly 35% of all students in American universities are taking at least one e-learning program (Allen and Seaman 2011). Another more recent phenomenon is the provision of online courses that are either provided in conjunction with elite universities such as Columbia, Brown, Princeton or Duke Universities for free, or by for-profit organisations such as Coursera founded in January 2011. According to Pappano (2012), these massive open online courses (MOOCs) are usually free and not for credit. They can be taken anywhere there is an internet connection. They are a strange amalgam of social networking, entertainment and networking. From this perspective, those for-profit courses offered by Coursera have reached 1.7 million users and this area is growing faster than Facebook (p. 15).

Other reports in 2012 indicated that corporate education was a $200 billion industry of which the portion of e-learning could be $56 billion and would be expected to increase to double by 2015. According to these reports it is expected that the market for e-learning will swell to $51.5 billion by 2016. From this perspective, it is expected that online courses and learning management systems alone would earn more than $7 billion by 2018 (McIntosh 2015).

According to the latest reports of e-learning status, while the rapid pace of online learning growth has moderated, it still accounted for nearly three-quarters of all US higher education's enrolment increases last year and the education system has strategic plans for the future of it (Allen and Seaman 2015), for example, Pennsylvania's State System of Higher Education will strive to achieve outcomes by 2020 which include increasing the number of students in online learning to 53,000 (PASSHE 2014).

Similarly, in Australian institutes, there are many statistics and reports in relation to e-learning growth rate which show that between 2009 and 2014 the online education industry in Australia experienced an annual growth of 14.4% with estimated revenue of over 6 billion dollars (IBS World 2014). Australian higher education's embrace of the use of e-learning as a vehicle to enhance teaching opportunities and improve learning outcomes is one of the strongest among developed countries in the globalization era. Open universities and distance learning institutions continue to offer students e-learning, using a diverse range of institutional policies to support the promised policies (Bates 1997). The providers and educational policy-makers are able to demonstrate that their processes in regard to online learning as a mode of delivery for their programs are sound and effective (Hosie et al. 2005; Oliver 2005). It can be concluded that adoption of online teaching and learning in the Australian higher education sector has been widespread and is now found across a range of disciplines (e.g., business, education, health, psychology, and accounting and information technology) and a range of program levels.

Because the main aim of this research is a comparative study of the e-learning practices in an Australian university and in a USA university, the researcher provides here some information of the e-learning maturity of the two universities.

The American university sample does not consider e-learning as simply a supportive technology of teaching and learning. Instead they regard it as critical to all educational provision namely concerning:

- relevance to all educational stages—undergraduate, graduate and professional, and continuing education.
- serving both resident and distance students.
- consisting of a wide range of approaches, from technology-enhanced classrooms and instruction to online courses and learning platforms.

The university's e-learning strategy is specifically directed at: (taken from e-learning at the American university sample, 2016):

1. Improving the undergraduate teaching and learning experience by targeting selected programs and courses for enhancement or redesign.
2. Supporting increased graduation and retention rates by giving undergraduates additional scheduling flexibility through redesign of high demand classes into an online format.
3. Providing graduate and professional students with alternative access to select post-baccalaureate programs by offering them in an online or blended format.
4. Improving access to university continuing education and noncredit offerings for professionals and lifelong learners.
5. Exploring the potential of emergent technologies by offering a limited number of massively open online courses (MOOCs) to a national and international audience.

The university has long had a reputation for adopting new technologies in order to improve access to education for the wider community.

From 1915 through 1999, their Audio Visual Library Services (later University Film and Video) provided educational films and videos to classrooms throughout the United States.

In 1946, their university station KUOM aired learning programs for children home bound by the polio epidemic.

From 1987 through 2003, the American university sample produced "Health Talk and YOU," a call-in TV show staffed by university medical experts and students. Currently, the Academic Health Center publishes the Health Talk blog.

In 1996, the first online courses were offered from two of their campuses, Twin Cities and Crookston.

In the fall of 2006, the Moodle course management system was launched, enabling faculty to provide students with course materials, library resources, and the ability to electronically submit their assignments.

In the fall of 2007, active learning classrooms were piloted on campus. In the summer of 2010, the Science Teaching and Student Services building opened with another 10 such classrooms. At that time, this provided more such facilities than any other institution in the country.

In the fall of 2010, the College of Education and Human Development launched a mobile learning pilot initiative.

In the 2014–2015 academic year, 1538 online course sections were offered and 21,451 students were enrolled in these, some in more than one. This was responsible for a 6.7% increase in total enrollments (41,333) over the previous academic year. The American university sample's commitment to on-line learning is such that it offers over 40 online and blended degree and certificate programs ranging from public health and nursing to computer science and manufacturing management (Academic Affairs & Provost 2016).

By contrast, the Australian university sample does not exhibit a long history of commitment to online teaching and learning although it is making endeavours to catch up. The 2016–2020 plan is aiming to position the university used in this case study as the best university in Australia and a leading institution globally. As part of the plan, Educational Technology Incubator (ETI) will be expanded to extend e-learning capacity to create video, animation, visualisation and simulation for teaching purposes, and to support the development and assessment of new tools, technologies and strategic innovation projects. Furthermore the university has plans for a massive development program for open online courses (MOOCs) (Strategic Plan 2016).

Currently, the e-learning system works with the University community to develop integrated learning spaces and e-learning systems to enhance the student learning experience. These projects and developments involve collaboration between many departments of the university

- integration of enterprise technology infrastructure for award course programs and units of study with faculties
- academic development programs and staff training with Institute of Teaching and Learning and University ICT
- student support and resource development with the Learning Centres
- learning Space development with ICT, CIS and Student services
- campus planning with the office of the DVC (Strategic Management),
- campus Infrastructure Services and committee structures
- business Intelligence development through metrics development with the Office of Information and Planning
- mobile Resources development with ICT and Marketing and Communications.

Activities arising from these collaborations are overseen by the SEG (Education), SEG (Infrastructure and Finance), SEG (Curriculum and Course

Planning), SEG (University Services), and SEG (Alumni and Marketing) committees (Academic Affairs & Provost 2016).

The goal, through a planned sequence of ICT projects, S-eLearning, is to achieve a single, integrated, enterprise-level virtual learning environment including the development of a ‘virtual extended classroom’ for every unit of study. The Open Learning Environment is planned to support self directed on-demand access to a pool of learning resources for all students, as well as access to workshop-supported modular courses on topics of interest to students (Strategic Plan 2016).

1.1 Statement of the Problem

While there is growing demand for e-learning projects in developed countries, nevertheless the failures are many, one of the main reasons being the quality of the e-learning projects (Shailaja and Sridaran 2014). Failing e-learning projects such as Universities 21, UKeU, New York online University, and The Global University Alliance demonstrate that the success of e-learning practice directly depends on quality and requires understanding of the current environment and the context of the e-learning practice (Inglis 2005; McLoughlin and Visser 2003; Oliver 2005; Salmon 2005; Smith et al. 2009; Smith et al. 2009).

One of the main reasons for an Australia–USA comparison of e-learning is to attempt to assess the extent and quality of e-learning programs in both countries. The academic ranking of both countries’ University systems shows a dominance by American institutions. “American universities dominate world rankings, irrespective of which ranking system is used. For example, in 2014, The Times Higher Education World University Rankings, which ranks universities on teaching, research, knowledge transfer and international outlook, had American universities occupying 8 of the top 10 places. Only five Australian universities made it into the top 100, and only one made the top 50” (The United States vs. Australia 2016). According to a range of studies, the United States has strong performance relative to Australia in e-learning practices.

One such study by Wills (2012) focused on research on e-learning for university students. That study revealed that “… the development of role-based e-learning over the past 20 years in Australia using simple e-learning technologies such as email and online discussion forums was quite different to that in America. America, by contrast, compares this with emerging forms of the e-learning design which are adopting newer technologies.” (Wills 2012, p. 2). The use of e-learning and virtual environments as a platform dominated the samples collected by Wills (up till 2009) for US, in contrast to the simpler technologies used in his Australian samples.

This book identifies information relating to e-learning that will be of assistance to Australian universities to improve their existing e-learning programs. An earlier Australian study (Nayda and Rankin 2009) highlighted that collaboration between faculty and staff was one approach in addressing the need for staff development to use digital technologies effectively. Staff development was also needed to develop online assessments and to provide skills in monitoring the quality of online courses and teaching strategies (Smith et al. 2009). Walker et al. (2006) made recommendations to address the unique learning needs of the X and Y generations and stated that educators must look at ways to enhance the learning environment to fit the expectation of these students. Consequently it was felt that by examining the e-learning practice in a top American university it would enable best practice to be established in Australian institutions.

As explained above, a large number of studies has shown the effect of a lack of quality and the need for an understanding of the current environment and the context of e-learning practice on students engaging in online activities. For example, research in the US and Australia has shown that engaging in a variety of online programs, awareness and confidence of working with IT are limited in students (Downing et al. 2014). Indeed there is a lack of homogeneity in students' contexts and a potential "digital divide" between students (Downing et al. 2014). Further, while students do engage in online activities, the new generation of students tends to use a "snatch and grab" approach to information gathering, (Bennett et al. 2008, p. 781) and has "shallow, random and often passive interactions with text" (Coiro 2003, p. 458).

Furthermore, according to US department of education reports and a survey of Harvard's Online Open Courses, engaging in online learning and activities has decreased with students and educators because of the lack of attention to many elements of quality and the context of e-learning practice (Reeves and Pedulla 2011; Reich 2014; Smith et al. 2009). From this perspective, McGorry (2003) asks for more attention to be paid to the quality of e-learning practice in higher education. Further Zhao (2003) recommends that universities implement a quality assurance plan aimed specifically at e-learning programs.

To conclude, concern about quality and achievement outcomes (Heafner et al. 2015), a lack of attention to learning from others' experience and perspective (no standard comparison of success with other institutes' practice), a lack of assessment and links to competency measurement (Hills and Overton 2010), existing poor managerial practices and lack of evaluation (Van der Vyver et al. 2015) and too much emphasis on technological practice without thinking about cultural pedagogical practice are the main challenges for enhancing and assuring e-learning in higher education world-wide. Indeed enhancement of quality of practices rather than quantity development is the more pressing concern to educational leaders and policy makers (UKIBC 2015).

It is worth mentioning that quality improvement is conceived as a constant enhancing of the process, outputs and outcomes of e-learning. Indeed those aspects of an e-learning course or unit which would be recognised as valuable may need quality improvement most (Inglis 2008). Quality enhancement is also "more transformative and it requires a deliberate change process- including teaching and learning- that is directly concerned with adding value, improving quality and implementing transformational change" (Lomas 2004, p. 158).

To enhance the quality of e-learning and best practices several main issues have been identified as needing improvement and are listed here: usefulness, perceived ease of use, support, and e-learning self-efficacy issues (Weng et al. 2015, p. 188), technical and learning environment issues (Madsen 2003), cultural resistance, technology and lack of interaction (Newton 2007, p. 29), technical difficulties, lack of a sense of community, time constraints, and lack of complete understanding of course objectives (Antoine 2011, p. 34; Song et al. 2004), cultures and faculty that resist change (Forsyth et al. 2010), optimising technology use to enhance the quality of student learning (Krause et al. 2009), regular attendance, suitable technologies and infrastructure and completion of tasks and programs (Hensley and Goldsmith 2013).

To overcome issues and enhance the quality of e-learning, several programs have been established. The Quality Matters Program has established national benchmarks for e-learning practices and has become a nationally recognised institute to certify the quality of online learning programs in USA (Butcher and Wilson-Strydom 2013). Also the Australasian Council on Open, Distance and e-Learning (ACODE) has provided benchmarks for the quality of technology to enhance e-learning experience in an institution-wide policy, planning for institution-wide services and institution-wide support (Sankey 2014). The main purpose of benchmarking and such strategies is to support continuous quality improvement in institutes based on action plans. For instance Federation University Australia's e-learning Plan from 2015 to 2017 focuses on several key aspects concerned with striving for and achieving excellence in practice. From this point of view, the alignment of practices includes the Australian Skills Quality Authority (ASQA) and the Tertiary Education Standards and Quality Agency (TEQSA). Another initiative of this institution is alignment of the quality of education followed by external standards to establish and maintain quality to ensure fulfilment and reach best applicable practice (Devlin 2015).

The commitment to quality improvement of e-learning needs to be built into a university's cultural context and assessment of e-learning practice to ensure the university continues to change and adapt to the needs of its students.

Ehlers (2009) proposes culturally sensitive frameworks for assuring and enhancing quality in e-learning practice (Masoumi 2010). The claim is that e-learning programs should be relevant to the context of the culture in which they have been applied. To get more success from e-learning programs, developed countries like the United States of America and Australia attempt to investigate how to individualise characteristics, technology and contexts of their e-learning system (Anderson and Gronlund 2009).

Cultural context and cultural dimensions are essential aspects of e-learning systems that both directly and indirectly affect their quality (Edmundson 2003; Masoumi 2010). Therefore cultural factors can be seen as the foundation for furnishing improved e-learning systems that can modify the whole e-learning structure (Kujala and Lillrank 2004). Cultural aspects such as educational paradigms, origin of motivation, experimental values, value of errors, role of instructor, user activity, learner control, accommodation of individual differences and collaborative learning (see Edmondson 2004; Gamble 2009; Henderson 1996; Masoumi and Lindström 2012; Reeves and Harmon 1994; Washburn 2012) focus on the pedagogical context of e-practice (Reeves and Reeves 1997) which may be oriented towards either constructivism or instructivism.

Currently, higher education systems and learning environments are changing from delivery-centred to learner-centred and from “showing-telling” to “learning-by-doing”; therefore it seems that the constructivist approach forms a strong theory on which to base new learning environments more suitable for the twenty-first century and the cultural dimension (Bednar 1992; Duffy and Jonassen 2013; Duffy et al. 2012; Jacobson et al. 2005; Kala et al. 2010; Putnam and Borko 2000; Stacey 2012; Tobin 1993; Young and Paterson 2007). The constructivism perspective takes an individual approach to constructing knowledge and conceives learner–learner interaction with the natural world as a way to construct their knowledge instead of injecting information and basic literacy which is the practice of instructivism (Jonassen 1991). According to e-learning practice based on constructivism, the students, lecturers and providers are actively involved in the pedagogical process and use cognitive and social tools for problem solving and knowledge transfer (Kelsey 2007; Low 2007; Weeks et al. 2013; Woo and Kimmick 2000). Consequently socio-constructivism concepts are the foundation of e-learning technologies in developed countries (Bjekic et al. 2010).

The principles of constructivist educational theory have come to be central to e-practice (Nkonge and Gueldenzoph 2006) and the influence of constructivist thought on e-pedagogy has provided basic principles of constructivism learning theory (Doolittle 1999; Hein 1991) and e-teaching best practices (Alley and Jansak 2001; Hacker and Niederhauser 2000; Keeton 2004). E-practice focused on learning and teaching processes is based on a function for operational policies and practice standards for virtual learning environments (Kala et al. 2010). According to the evidence, the use of practice based on learning and teaching theories can support online learning courses by developing a model for the learning and teaching process (Oliver 2001; Thurmond 2002).

Further, the comparative approach to cultural issues is one of the main factors for enhancement of quality improvement of e-learning practices (Adamson 2012; Alexander et al. 1999; Arnove et al. 2012; Bignold and Gayton 2009; Bray et al. 2007; Kubow and Fossum 2007; Thomas 1993; Wolhuter 2008). This approach can raise awareness of the differences and make clear the similarities between systems and practices in various countries (ibid.). Comparative study approaches have been shown to be predominantly evidenced-based and justified as frameworks of e-learning (Adamson 2012; Alexander et al. 1999; Arnove et al. 2012; Bignold and Gayton 2009; Bray et al. 2007; Kubow and Fossum 2007; Thomas 1993; Wolhuter 2008).

Therefore given the significance of cultural context in e-learning and the quality improvement of its practice, investigation of comparative dominant cultural dimensions of e-learning practices and assessing e-learning practices such as pedagogical, performance appraisal, instructional design, technological, administrative and support service practice in different cultural contexts are critically important (Chickering et al. 1987; Commissions 2001; Dragon et al. 2013; Finger et al. 2006; FitzPatrick 2012; Holsapple and Lee-Post 2006; Kala et al. 2010; Khan and Granato

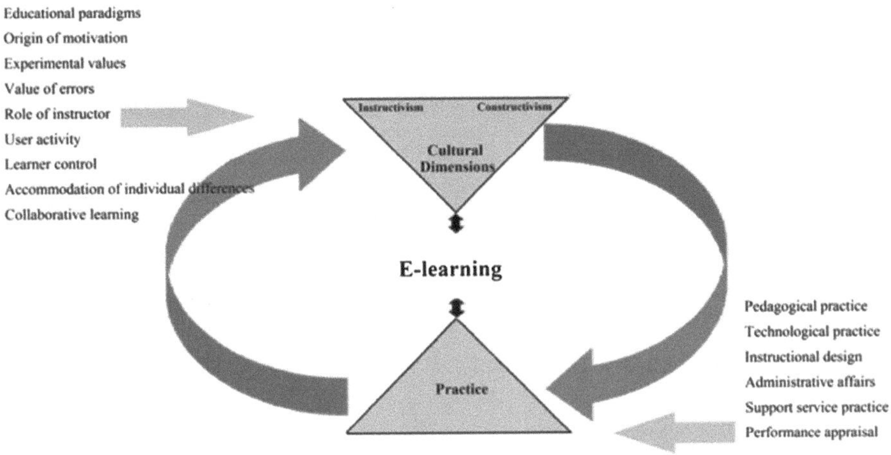

Fig. 1.1 Theoretical framework

2008; Marshall 2012; Sangrà 2002; Selim 2007; United States Institute for Higher Education Policy 2000; Zhao 2003; Zhou 2012) (see Fig. 1.1).

From this perspective, comparison of e-learning practices between Australian and United States institutions with a view to the restructure of higher education would present benefits such as (a) recognition of best e-learning practices through examining national and international experiences and (b) application of best national and international e-learning practices provided by advanced technology to obtain positive effects on experiences, strategies and approaches involved (VET 2012).

To conclude, with due attention to fast-growing e-learning programs in the institutional and pedagogical structures, there is no doubt that comparative studies on virtual learning environments will lead to fundamental change in the educational process, because focusing on a variety of opinions and experiences in different systems and cultures would lead to the identification of strategic issues (strength, weakness, opportunity, and threat). Also "the use of comparative studies has become a prominent feature in policymaking and related processes which is characterised by increased technological, information and pedagogical transfer" (Adamson 2012, p. 641).

1.2 Purposes of the Study

Despite the large number of researchers focusing on e-learning issues in recent years, there is still limited knowledge about many different issues concerned with e-learning practices in a comparative context. Consequently, the main aim of this research is a comparative study of the dominant cultural dimensions, e-learning practices and current issues and problems between one faculty in an Australian university and one faculty in an American university. This will ultimately improve the quality of online learning courses in an Australian university and an American university. To reach this aim the researcher asked three main questions, namely:

1. What are the dominant cultural dimensions of e-learning practice in an Australian and an American university?
2. What is the current status of e-learning practices in an Australian and an American university?
3. What are the dominant issues of e-learning practices in an Australian and an American university?

It is hoped answers to these question will illuminate 2 main characteristics: cultural aspects and e-practice factors.

1.3 The Process and Structure of the Book

The research process of this study involved firstly, identifying the relevant factors of cultural dimensions and the aspects of best practice of e-learning by reviewing relevant documents and providing a suitable framework. Secondly, three comparative studies between an Australian university and an American university were conducted aligned with the three research questions. The results of each study were obtained and subsequently a comparison of the results of each individual factor within the American and Australian samples was made and both are presented and discussed.

The first study is concerned with cultural aspects of Australia and the United States of America. The two dimensions of objectivism-instructivism and socio constructivism are used to help show how these cultural dimensions can be classified within the broader definition of those terms. The second study looked at e-practice aspects of the two countries namely pedagogical, performance appraisal, instructional design, technological, administrative and support service practice (see Chap. 2). Two quantitative surveys were associated with the first and second research questions. The third study addressed the current problems of the two countries' case studies particularly the pedagogical, technological and cultural aspects. To answer the third question of research a qualitative approach is used, firmly based within a comparative framework.

The book is organized into six chapters. An overview, stated problem, aim and questions of the book are provided in the first chapter. The second chapter consists of three parts; the first part provides an outline of the context of the study. The aim is to give a brief account of the rapidly expanding e-institutions in the light of reality, progress and difficulties by looking at the historical situation of e-learning. The second part examines cultural issues in the e-learning practice. Initially, it addresses common educational paradigms and then gives an outline of cultural dimensions in educational settings. A cultural model in e-learning practice is mapped out. In the third part, e-learning and relevant theoretical frameworks, which we need to consider before moving forward, are addressed. Specifically, theoretical and practical research studies in the e-learning field based on comparative characteristics are described in order to develop an e-practice model in online learning courses. Best practice and models up to the present are reviewed based on the comparative approach in university environments.

The third chapter of this book presents the dominant cultural orientations of e-practice in Australian and American e-learning systems and then compares the results of the two countries. The fourth chapter describes the current status of Australian and American e-learning practices. The fifth chapter poses the current problems of the two countries. Current problems are discussed in relation to the pedagogical, technological and cultural aspects.

In the beginning of the third, fourth and fifth chapters the research method are discussed. In addition the design and procedures of data collection, descriptive information of participants, and the analytical procedures are explained. Finally, the sixth chapter discusses the findings of all three studies, and presents the conclusions limitations and recommendations for further studies.

Chapter 2
Review of Related Literature

2.1 Introduction

This chapter presents a review of the literature and consists of three parts. The first part traces, via various studies, the historical background and rapid development of e-learning and presents some current definitions of the concept. The second part investigates the literature in the e-learning field that focuses on cultural issues. An overview of the cultural dimensions in educational settings in general and in virtual institutions in particular is presented. In the third part, relevant e-learning theoretical frameworks, which we need to consider before moving forward, are addressed. Specifically, theoretical and practical research studies in the e-learning field based on comparative characteristics are described in order to develop an e-practice model in online learning courses. Best practice and models up to the present are reviewed based on the theoretical and practical research in university environments.

2.2 Historical Background

E-learning began as distance education so, historically, e-learning and distance education have been linked together in the literature (Bowles 2004). Approximately, distance education has a history that spans two centuries (Spector et al. 2008) when the system of correspondence education was developed in order to reach a geographically dispersed population (Moore and Kearsley 1996). Distance education encompasses a variety of delivery media, learning technologies and online instruction (Allen and Seaman 2007; Keegan 1990). It can be argued that development of distance education has been linked with the emergence of new technological tools and styles (Chaney 2004). The development of distance education can be articulated in terms of various historical developments as indicated in Fig. 2.1 (Breitner and Hoppe 2005).

S.H. Sadeghi, *E-Learning Practice in Higher Education: A Mixed-Method Comparative Analysis*, Studies in Systems, Decision and Control 122,
https://doi.org/10.1007/978-3-319-65939-8_2

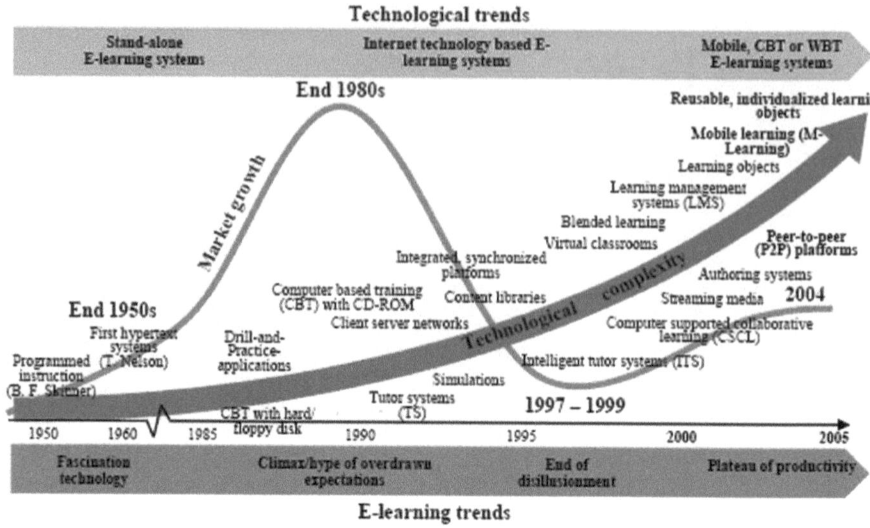

Fig. 2.1 Distance learning trends

2.2.1 *First Generation: The Correspondence Model*

This era is regarded as the first generation of distance education, primarily using printed media and having an effect on the pedagogical structure of instruction. It is known as the Correspondence Model (Sloman 2002). This generation is marked by the linear delivery of knowledge and information beyond geographical barriers where learning and teaching is organized asynchronously (Bramble and Panda 2008).

2.2.2 *Second Generation: The Multi-media Model*

Radio and television were the emerging technologies in the distance education field in this era. During the 1950s, Kansas State College, Purdue University, and Iowa University piloted TV programming and Production for distance education (Buckland and Dye 1991). This era began with the widespread use of radio and television as the carriers and presenters of the courses in addition to printed material. These media required different teaching and learning strategies and technics, which changed the pedagogical structures of the courses (Taylor 1999).

2.2.3 *Third Generation: The Tele-Learning Model*

With the increased accessibility to television and videoconferencing in the 1980s, many postsecondary schools began offering videoconference courses between different campuses (Hancock 1999). These media created a new pedagogical model to deal with the new central medium of videoconferencing (Sloman 2002).

2.2.4 *Fourth Generation: The Flexible Learning Model*

According to Im (2006), this model involves more fully-fledged capacities of World Wide Web technologies which provide a three-dimensional e-learning model (anytime, anywhere at any pace). The use of the Internet in online learning provided the opportunity for the student to experience a learning process that is collaborative, nonlinear, and interactive (Taylor 1999). Students were now able to learn anywhere, anytime and at any pace, encouraging social interaction such as collaborative learning and Computer Supported Collaborative Learning (CSCL), and situated learning (Bramble and Panda 2008).

2.2.5 *Fifth Generation: The Intelligent Flexible Learning Model*

The still emerging Fifth Generation model of distance education, which is known as the Intelligent Flexible Learning Model, is based on on-line delivery systems from the Internet (Taylor 1999). According to Taylor (1999), "The fifth generation of distance education is essentially a derivation of the fourth generation, which aims to capitalize on the features of the Internet and the Web" (Taylor 1999, p. 2). This model has the potential to provide learners with a much more personalized educational experience compared to the prior generations of distance education. Currently this generation is identified as involving various blended learning approaches (Taylor 2010).

2.3 E-Learning Definition

The rapid growth of e-learning over recent decades has spawned many definitions that address different features of the e-learning concept, but most of these definitions focus on the same set of features. Some are basic, e.g. E-Learning includes the use of the online tools and other technologies to produce materials for education, to teach learners, and also to regulate courses in an organization (Fry 2001). Others claim the e-learning concept describes a variety of different forms of use of digital technologies in pedagogical processes (Fischer et al. 2015). Some scholars define e-Learning as any educational offering that makes use of ICT for asynchronous, decentralized content presentation and distribution, as well as for interpersonal communication and interaction (Holmberg et al. 2005).

Some authors argue that e-learning refers to the key frameworks of learning and teaching that are enabled in some way by ICT tools to deliver a broad array of solutions with the purpose of enhancing knowledge and performance (Mason and Rennie 2006). In a similar way Stockely (2003) argues that e-learning involves the use of a PC in some way to deliver learning and teaching material. Some of the definitions focus on the electronic tools and applications used in the processes of learning and teaching in a dynamic and intellectually challenging learning environment (Garrison 2011).

As explained in the definitions above, "electronic learning" or e-learning presents a wide range of online technologies to produce materials for enhancing knowledge and performance of education.

A number of these definitions focus on social networking communication and some on the electronic tools and applications used in the processes of learning and teaching in a dynamic environment. For example:

The Open and Distance Learning Quality Council in the UK presents e-learning as 'the effective learning process created by combining digitally delivered content with support and services" (Masoumi 2010, p. 46).

"E-learning is learning based on information and communication technologies with pedagogical interaction between students and the content, students and the instructors or among students through the web" (González-Videgaray 2007; Sangrà et al. 2012, p. 3).

Ministry of Communication and Technology of New Zealand presents e-learning as "… learning facilitated by the use of digital tools and content that involves some form of interactivity, which may include online interaction between the learner and their teacher or peers" (Sangrà et al. 2012, p. 3).

In 2015, Online Learning Consortium defines e-learning based on two key characteristics; "they include definitions at both the course level and the program level" and also "they incorporate three key parameters: instructional delivery mode, time, and flexibility" (Mayadas et al. 2016, p. 3). These characteristics define e-learning as a new paradigm based on pedagogical oriented definitions. In this point of view e-learning refers to "pedagogical processes that utilise ICT to mediate both synchronous and asynchronous learning and teaching practices" (Jereb and Šmitek 2006). Therefore e-learning is defined as ICT, computers and networks used to support learners to improve their learning and educational processes (Ellis et al. 2009).

Many e-learning definitions are more focused on the instruments used in the new environment in the digital age which creates student-centred learning and educational practice, offering new more flexible learning methods (Shopova 2012). Because some of the definitions focus on the technology base rather than pedagogy, it has been stated that "describing e-learning in terms of the enabling technologies is not useful as this does not distinguish between the types of design features for

Table 2.1 Pedagogical dimensions of type I and type II e-learning

Pedagogical dimension	Type I e-learning delivery-centered paradigm	Type II e-learning learner-centered paradigm
Learning mechanism	Showing and telling	Learning-by-doing
Role of technology	Delivery of content and evaluations	Cognitive tools, scaffolding learning, providing feedback, non-linear access to information sources, and supporting collaboration
Type of content	Didactic written texts and multimedia lectures covering factual information	Realistic texts and multimedia cases, 2D and 3D simulations and virtual worlds, dynamic computer models
Role of student	Passively view or receive content	Actively engaged in problem solving, projects, and collaborative activities
Control	Technology	Learner
Learning outcomes	Achievement on objective memory tests of factual information retention	Ability to solve new problems, performance assessments

(Jacobson et al. 2005)

various e-learning approaches, and more important, between different paradigms for teaching and learning" (Jacobson et al. 2005, p. 79). Consequently, it is important that the definitions of e-learning cover a broad range of pedagogical dimensions.

Jacobson et al. (2005) proposed pedagogical dimensions based on types of e-learning: type I e-Learning is a delivery-centred paradigm (more traditional approaches) and type II e-Learning is a learner-centred paradigm (active engagement in doing activities). The pedagogical dimensions of Type I and Type II e-learning are shown in Table 2.1

As can be seen in this table, Type I differs from Type II e-learning in a number of dimensions. The pedagogical dimensions of type I focus on learning basic factual information. However, the pedagogical dimensions of type II focus on constructing deep understandings of the content and problem solving involving learning-by-doing (Jacobson and Spiro 1994). Type II e-learning systems represent advanced research efforts that are exploring ways to design and use advanced technologies to support learner centred pedagogies (Jacobson et al. 2005).

2.4 Cultural Context in E-Learning

As one of the major foci of this study is the impact of cultural factors on e-learning, this section will review the literature relevant to this aspect. Many variables within a particular cultural context can impact the design of e-learning such as learner values, student perceptions, styles of communication, and desired learning outcomes (McLoughlin 1999). In fact because cultural context is an inevitable aspect of e-learning systems, it both directly and indirectly affects their quality (Edmundson 2003; Masoumi 2010). Therefore cultural factors can be seen as the foundation for furnishing improved e-learning systems that can modify the whole building of an online learning environment (Kujala and Lillrank 2004).

According to Seufert (2002), the cultural impact on e-learning can be viewed from two core perspectives: design and use.

As shown in Fig. 2.2, the cultural context impacts on e-learning environments in different ways particularly on the areas of Designing and Using. Its impact on design is initially by imposing learning paradigms, organizing resources and so on. It also has an impact on the use of online learning in terms of and use of technological acceptance, learning and teaching style and scenarios, instructor behaviours (Adelsberger et al. 2013).

Rogers et al. (2007) offered the concept of building bridges into e-learning as a way that allows for flexibility. Bridges can create an awareness and flexibility in areas such as relationships and social and educational expectations. However, the key to being able to build bridges into e-learning involves an initial understanding of the key differences in the expectations and abilities of learners from different cultures. It is believed that e-learning environments cannot be transferred in an isolated manner without the culture-related roots context in which they are produced and created (Wild and Henderson 1997).

A large number of studies (cf. Butter et al. 2015; Chen et al. 2012; Collis 1999; Edmundson 2004; Henderson 1996; Masoumi 2010; Reeves 1994; Usun 2004) has investigated the importance of cultural issues in online learning. In fact, many mainstream researchers, such as Reeves (1994) and Henderson (1996) have focused specifically on designing models for assessing the cultural dimensions of education and educational artefacts, including e-learning.

Henderson (1996) identified a need for understanding different cultural learning requirements in regard to instructional design. Based on Reeves's (1992) model, Henderson (1996) developed a comprehensive Multiple Cultural Model for investigating the cultural aspects in e-learning environments.

The MCM contains 14 dimensions of learning, depicted by extremes separated by double arrows in Fig. 2.3. The concept of this model is similar in design to those used

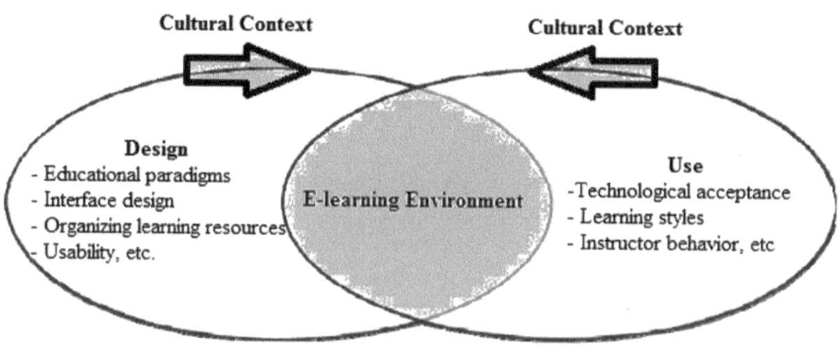

Fig. 2.2 The cultural impact on e-learning

by Hofstede (2001) and Trompenaars and Hampden-Turner (1998). Henderson (1996) believed these 14 dimensions are related and all impact the design of e-learning. Basically, this system is a structure of "eclectic paradigm," which requires designing materials and resources that allow flexibility while enabling the learner to learn via interaction and collaboration with resources that reflect multiple cultural perceptions, including multiple ways of learning and teaching. It promotes equity of learning outcomes via combining and adapting mainstream and non-mainstream cultural interests (Gunawardena and Wilson 2003).

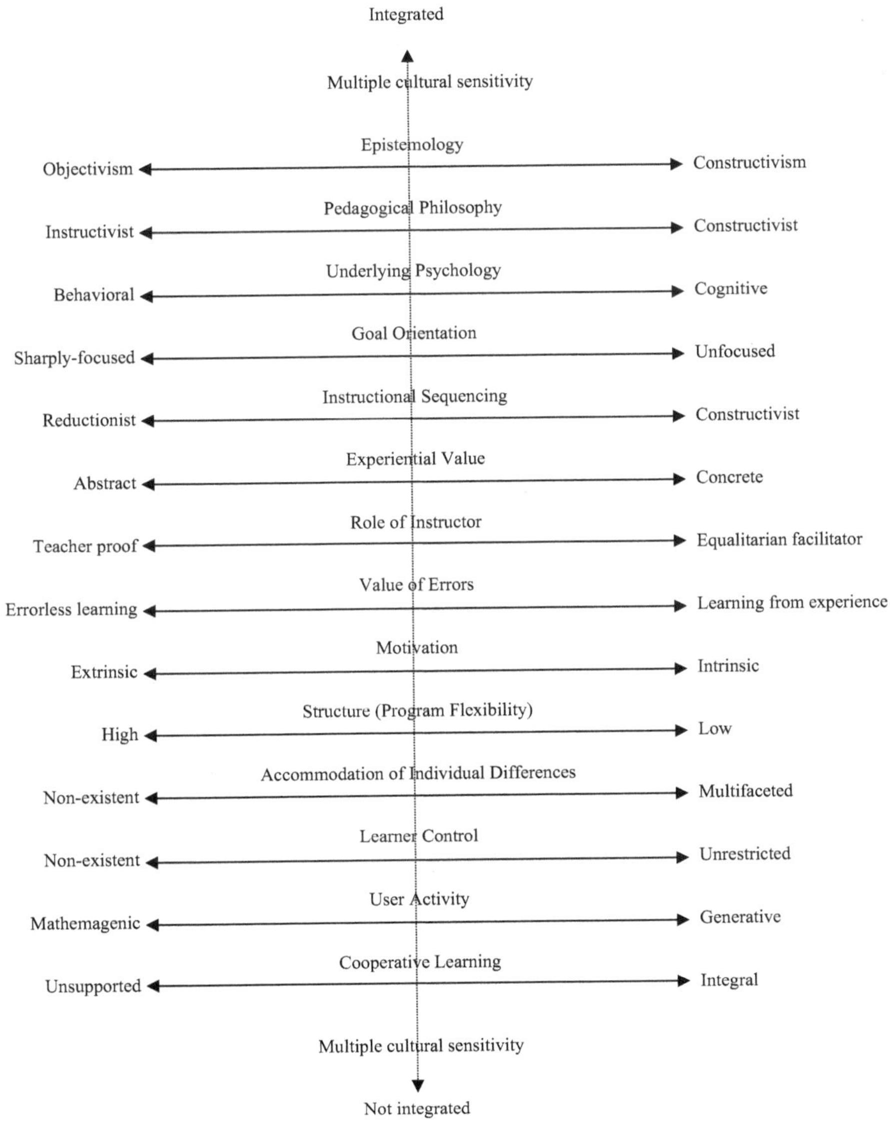

Fig. 2.3 Henderson's multiple cultural model

Using Henderson's (1996) model, Edmonson (2004) focused on comparing India and America, Gamble (2009) compared China and America, Morris (2009) and Washburn (2012) analysed Asian and American e-learning systems and Masoumi (2010) focused on the quality of e-learning in an Iranian cultural context.

A study on the cross-cultural dimensions of globalized e-learning conducted by Edmundson (2004) is of particular interest. Edmundson's study reviewed data from India and the United States in an effort to find differences or similarities between learning outcomes, learning preferences, and cross-cultural dimensions related to e-learning. Edmundson used the MCM by Henderson as the foundation for a questionnaire to generate an understanding of the learner's preferences and perceptions that could help e-learning adapt to a multicultural audience. However, some of the dimensions in the original MCM entail relatively similar cultural constructs. For instance, the three dimensions, epistemology, pedagogical philosophy and underlying psychology, address the underlying educational paradigms (Masoumi 2010).

Consequently, a simplified MCM was developed by Edmundson because she believed that the strong interrelationships between the first four dimensions of the original MCM represented multiple learning paradigms that could be represented as what Edmundson titled a "pedagogical paradigm" (Edmundson 2004, p. 123).

As in the original MCM depicted by Henderson (see Fig. 2.3), the objectivist–instructivist paradigm is on the left and the constructivist–cognitive paradigm is on the right. The SMCM is depicted in Fig. 2.4. It should also be noted that there may be other cultural dimensions that still need to be delineated (Reeves 1992).

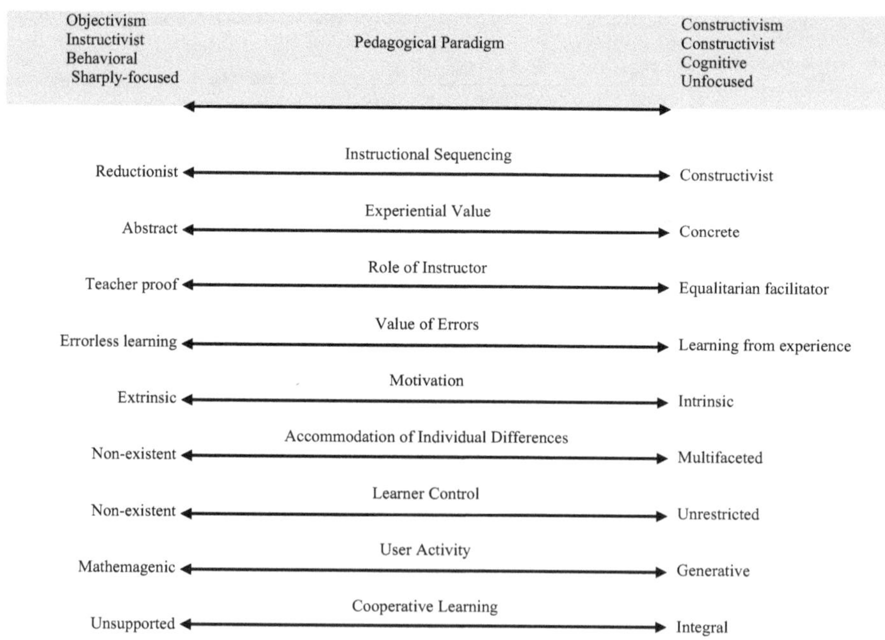

Fig. 2.4 A simplified multiple cultural model by Edmundson (2004)

These models, that map out the cultural orientations in e-learning contexts, provide a valuable framework for evaluating and judging an educational setting and provide a profile of the e-learning environment. Consequently, each of the cultural dimensions they list, which may be either constructivist or instructivist, will be given a brief overview below.

2.4.1 Pedagogical Paradigm (Objectivism vs. Constructivism)

The Pedagogical Paradigm is concerned with theoretical enquiries into the nature of knowledge and teaching. The Pedagogical Paradigm also includes underlying educational paradigms that are embedded in the cultural context such as epistemology, pedagogical philosophy and psychology (Masoumi 2010). Understood broadly, epistemology is concerned with the creation, nature, dissemination and limits of human knowledge and so is an essential aspect of pedagogy (Scardamalia and Bereiter 2006).

Henderson (1996) offered objectivism (instructivism) and constructivism as opposites due to the nature of their contrasting assumptions. Objectivist epistemology establishes a definitive and transformative structure of knowledge, as opposed to constructivism, which calls for a variety of perceptions so that students can construct their own structure of knowledge and way of discovering and thinking. Similarly, in the instructivism/objectivism approach, goals and objectives are considered apart from the learner and focus on direct instruction. Assessments/evaluations are conducted concretely on the foundation of the given goals. In other words instructivists have an opinion on accumulation of knowledge and the role of the instructor to facilitate passing that knowledge and skill through clear goals and objectives (Rezaei and Katz 2002). However, the focus in constructivism is on the goals and needs of the learners, along with their previous experience and their meta-cognitive strategies (Reeves 1994, pp. 223–237). In fact constructivists believe students build and create new knowledge from previous knowledge (Huang 2002).

2.4.2 Goal Orientation (Knowledge Acquisition and Sharply Focused vs. Knowledge Transfer and Unfocused)

The objectives of a online course can range from focused and knowledge acquisition (focused) to knowledge transfer (unfocused), depending on the goals of the course and the expectations of teachers (Masoumi 2010). Focused and knowledge acquisition strategies emphasize clearly delineated goals, while unfocused strategies emphasize self-discovery and knowledge transfer (Edmundson 2004).

2.4.3 Experiential Value (Abstract vs. Concrete)

The Experiential Value aspect ranges on a continuum from abstract to concrete (Edmundson 2004). Ndoye (2003) and Kolb (1984) pointed out the importance of learning through practical, contextualized, learning situations with hands-on learning experiences. Abstract learning emphasizes the value of theoretical knowledge while concrete learning encourages real-life experiences (Washburn 2012).

2.4.4 Instructor's Role (Didactic vs. Facilitative)

This aspect is placed on a continuum from didactic to facilitative. The teacher's role may vary from being a facilitator of e-learning environments to being a transmitter and source of knowledge. The teacher's "didactic role" in a learning state may strongly scaffold the students and learning activities. Correspondingly, the learning audience, independent activity and practice may be increased when the lecturer stays in the background of the learning environment, as a facilitator (Reeves 1994). A didactic exposition of learning contrasts with pedagogical techniques that facilitate exploratory learning without controlling outcomes (Edmundson 2004). A learner that prefers the lecturer to provide the knowledge and believes an instructor should be an expert on the subject matter would be considered to have a preference for didactic teaching (Smerdon et al. 1999). However, the didactic process is instructor-centred learning and does not place importance on the student's previous experiences (Smerdon et al. 1999). The facilitative instructor assists students to create knowledge from previous experiences, encourages goal setting, creates various teaching methods and styles, promotes self-regulated and self-directed learning, and provides continuous response (Holly et al. 2008).

2.4.5 Program Flexibility or Structure (Teacher-Proof vs. Easily Modifiable)

A teacher-proof program restricts instructors from altering or changing instruction, while an easily modifiable program allows instructors to adapt instructions to the needs of the students (Edmundson 2004). An instructor-proof curriculum does not allow flexibility or varied adaptations. However, an easily modifiable instructional process allows flexibility when needed for increased learning and effectiveness such as by varied learning methods, lectures, experiments, inquiry learning, field trips, and authentic assessment (Reeves 1994).

2.4.6 *Value of Errors (Errorless Learning vs. Learning from Experience)*

Under an errorless learning pattern, a learner learns until he/she generates no mistakes. This instructional technique does not allow for errors. However, the learning based on the experience approach to instruction uses errors as part of the learning and teaching process (Edmundson 2004). Errorless learning refers to avoiding and eliminating incorrect responses and also encourages reducing the errors while learning and discovering (Mueller et al. 2007). However, lecturers who believe in the learning with experience approach encourage learners to learn from their errors as a process of learning and studying (Reeves and Reeves 1997).

2.4.7 *Origin of Motivation (Extrinsic vs. Intrinsic)*

The "source of motivation" aspect ranges from extrinsic (outside the learning environment and learners) to intrinsic (integral to the learning environment and learners) (Masoumi 2010). Extrinsic motivation stimulates based on aspects outside the individual (Merriam and Caffarella 1999). However, "intrinsic motivation originates from within in regard to particular academic tasks" (Walker et al. 2006, p. 4). Although intrinsic motivation and the love of learning are precisely what attract many learning audiences to learn in learning environments, intrinsic motivation mostly takes a backseat to the extrinsic motivational elements that learners consider important (Reeves and Laffey 1999).

2.4.8 *Accommodation of Individual Differences (Non-existent vs. Multifaceted)*

The impact of individual differences is a very critical factor in the effectiveness of educational settings (Masoumi 2010). This dimension can be dismissed as non-existent or embraced as multifaceted. In some learning environments there is no need for accommodation of individual differences because learning and knowledge are structured. However, when accommodation of individual differences does exist, knowledge and learning are presented in a variety of ways so that students can utilize the instruments that most suit their preferences and priorities (Edmundson 2004). An instructor using a multifaceted accommodation curriculum recognizes the different learning attitudes, previous knowledge, experiences, motivations, cognitive styles, and learning styles of students. The instructor would also acknowledge and accommodate the ways each individual accepts, processes, organizes, and retrieves information. While many instructors acknowledge and accommodate the multifaceted instructional process, others do not believe in accommodating individual differences. Scaffolding and metacognitive strategies are two practical methods to accommodate individual differences using multifaceted practices (Edmundson 2003; Reeves and Reeves 1997; Rosenfeld and Rosenfeld 2004).

2.4.9 *Learner Control (Non-existent vs. Unrestricted)*

In this dimension, the student must either learn along a predetermined path or learn by discovery. In other words, the student has either partial or complete control of the learning way (Edmundson 2004). An instructor who considers learner control is non-existent believes that learners achieve better feedback and performance with greater degrees of learning control, therefore the lecturer dictates the students' entire learning process (Washburn 2012). However, less supervision is required by an unrestricted student-control instructor, as the students establish their own way, process and decisions about their learning (Chou and Liu 2005).

2.4.10 *User Activity (Mathemagenic vs. Generative)*

A mathemagenic approach directs students to access the same material in different paths. A generative strategy encourages students to engage in the process of discovering, creating and elaborating (Edmundson 2004). Mathemagenic learning environments only offer activities that are relevant to specified lecturer-designated goals, and specific statuses (Rothkopf 1970). However, generative learning emphasizes the students' control of their educations through creating and engagement (Reeves 1994).

2.4.11 *Cooperative Learning (Collaborative Learning vs. Unsupported Learning)*

Learners can work independently and unsupported or they can be encouraged to participate via cooperative activities and practices (Gamble 2009). Collaborative learning refers to methodologies and environments in which learners discuss and share concepts related to the problems arising (Dillenbourg 1999). Gokhale (1995) described collaborative learning as "an instruction method in which students at various performance levels work together in small groups toward common goals" (p. 1). However, unsupported learning is believed to encourage individual development, including especially critical and creative thinking, and problem-solving. Elias (1997) explained how individual learning moves on a continuum and process "toward self-agency and authorship" as learners "increasingly recognize their responsibility for their actions, choices, and values for the decisions they may make based on those values" (p. 163). In some environments, there is no supportive collaborative learning system and in others, it is integrated in learning and teaching environments. There is a considerable body of research supporting the benefits of collaborative learning and cooperative learning (Reeves and Laffey 1999).

To sum up, from this overview of the main cultural educational paradigms, it is apparent that the underlying educational sitting in online environments not only forms part of the design, but also shapes the ways and approaches that these environments could be improved and developed. Further, by reflecting the contrast between objectivism and constructivism, these dimensions characterize both values and practices in educational settings. In the same vein, a number of scholars have argued that the most important element of a shift from traditional education to online is to acknowledge the change in cultural aspects such as the nature of the tasks, the lecturer's role in terms of didactic or facilitative (cf. Hase and Ellis 2001; Reeves and Reeves 1997; Wang and Reeves 2006).

Finally the term "culture" as used in this research impacts on e-learning environments in different ways particularly on the areas of Designing and Using. As can be seen in Fig. 2.1 of page 24, according to Adelsberger et al. (2013), its impact on design and settings as educational culture is initially by imposing learning paradigms, organizing resources and so on. It also has an impact on the use of online learning in terms of and use of technological acceptance, learning and teaching style and scenarios, instructor behaviours as cultural issues and sensitivities. "Culture" on the areas of designing and educational siting is a part of the quantitative study while "culture" in "Using" and "Cultural issues and Sensitivities" is part of the qualitative study in this research.

The next section reviews some theoretical frameworks proposed by various scholars and educational institutions on which to base best e-learning practice.

2.5 E-Learning Practice

The goal of most researchers in the e-learning field is to provide a comprehensive pattern of practice that can be applied to all students and all learning environments based on the success and quality of distance education and e-learning (Boud and Molloy 2013). While cognitive and social tools are considered as the main factors of effective education, the quality of both the pedagogical and technological content is also viewed as an important issue influencing learning practice (Ali et al. 2004; Cottrell and Donaldson 2013; Gerjets and Hesse 2004; Hiemstra 2009; Kala et al. 2010; Wilkinson et al. 2004).

According to the evidence, practice based on learning and teaching theories can support online courses by developing a model for the learning and teaching process (Oliver 2001; Thurmond 2002). Although several approaches and learning theories have been recognized in the e-learning field including behaviourism (based on observable behaviour-objectivity), cognitivism (based on unobservable behaviour-subjectivity) and constructivism (based on construction of new knowledge by doing, problem-based learning, conceptual understanding, problem solving and knowledge transfer), currently, the higher education systems and learning environments are changing from delivery-centred to learner-centred and from 'showing–telling' to 'learning-by-doing', therefore constructivist approaches form a strong theory in new learning environments for the 21st century (Bednar 1992; Duffy and Jonassen 2013; Duffy et al. 2012; Jacobson et al. 2005; Kala et al. 2010; Putnam and Borko 2000; Stacey 2002; Tobin 1993; Young and Paterson 2007). According to e-learning practice based on constructivism, the students, lecturers and providers are actively involved in the pedagogical process and use the cognitive and social tools for problem solving and knowledge transfer (Kelsey 2007; Low 2007; Weeks et al. 2013; Woo and Kimmick 2000).

Accordingly, a number of e-practice models based on constructivism in online learning courses that might improve the condition of online learning courses in Australian and American universities are reviewed below.

2.5.1 *Principles for Good Practice in (1987)*

The American Association of Higher Education (AAHE) provided some Principles for Good Practice in 1987 (Chickering and Gamson 1987). Comparing the models of, and frameworks for, practice discussed in more recent studies, it is still valid as a benchmark for the e-learning field (Partridge et al. 2011). The seven practices outlined are:

1. Encourage contact between students and faculty; the Internet, e-mail and learning management systems.
2. Develop reciprocity and cooperation among students; co-operative learning online.
3. Use active learning techniques; communication tools, online activities, electronic portfolios.
4. Give prompt feedback; e-mail, online discussion forum.
5. Emphasize time on task; asynchronous access and computer record keeping of time spent.
6. Communicate high expectations; real life problems and scenarios, public scrutiny of work submitted.
7. Respect diverse talents and ways of learning; variety of learning experiences, anywhere, anytime learning.

2.5.2 *Benchmarks for Success in Internet-Based Education in the US (2000)*

In 2000 the Institute for Higher Education Policy in the United States developed Benchmarks for Success in Internet-Based distance learning. The IHEP research is a comprehensive study of guidelines for success in e-learning environments and blended learning that include seven categories and 24 benchmarks: (1) institutional support; (2) course development; (3) teaching/learning; (4) course structure; (5) student support; (6) faculty support; and (7) evaluation and assessment (Phipps and Merisotis 2000). A large number of studies across the world have used this model as a basic framework (e.g. Yeung 2002; Herman 2001).

- Institutional support
  1. A documented technology plan.
  2. The reliability of the technology delivery system is as failsafe as possible.
  3. A centralized system provides support for building and maintaining the distance learning infrastructure.
- Course development
  1. Guidelines regarding minimum standards are used for course development, design.
  2. Instructional materials are reviewed periodically.
  3. Courses are designed to require students to engage themselves in learning activities.
- Teaching and learning
  1. Student interaction with faculty and other students.
  2. Feedback to student assignments and questions is constructive and provided in a timely manner.
  3. Students are instructed in the proper methods of effective research, including assessment of the validity of resources.
- Course structure
  1. Before starting an online program, students are advised about the program to determine self-motivation and course design.
  2. Learning outcomes for each course should be clearly written.
  3. Students have access to sufficient library resources.
  4. Faculty and students agree upon expectations regarding times.
- Student support
  1. Students receive information about programs.
  2. Students are provided with hands-on training to aid them in securing material.

3. Throughout the duration of the course students have access to IT assistance.
4. Questions directed to student service personnel are answered accurately.

- Faculty support

1. Technical assistance in course development is available to faculty.
2. Faculty members are assisted in the transition from classroom teaching to online instruction and are assessed during the process.
3. Instructor training and assistance, including peer mentoring, continues through the progression of the online course.
4. Faculty members are provided with written resources to deal with issues arising from student use of electronically-accessed data.

- Assessment

1. The program's educational effectiveness and teaching/learning process is assessed through an evaluation process that uses several methods and applies specific standards.
2. Data on enrolment, costs, and successful/innovative uses of technology are used to evaluate program effectiveness.
3. Intended learning outcomes are reviewed regularly to ensure clarity, utility, and appropriateness.

2.5.3 The Learning Sciences Framework for E-Learning Systems (2001)

Through an extensive review of related literature, Jacobson et al. (2005) attempted to develop a model for advanced E-Learning Systems in the Korean Association of Multimedia-Assisted Language Learning. This project is related to Learning Sciences Based Design Principles that are provided by the US National Science Council to study what students know. The eight principles of this model are (see Jacobson et al. 2005; Magoulas and Chen 2006):

1. Provide Contextualized Learning
2. Provide Scaffolds and Tools
3. Consider Students' Preconceptions and Prior Knowledge
4. Make Organizing Conceptual Structure Explicit
5. Provide Formative and Summative Assessment of Learning
6. Foster Thoughtful Collaborations amongst Communities of Learners
7. Support Learning-By-Doing
8. Provide assessment of learning using measures of knowledge transfer (Putnam and Borko 2000).

The pedagogical dimension in various types of e-learning is one of Jacobson's main concerns and includes learning the mechanism, control, role of technology, role of student, type of content and learning outcome (Jacobson et al. 2005).

2.5.4 *A Benchmark for European Virtual Campuses (2002)*

In 2002 European Commission DG Education and Culture provided a benchmark for European virtual campuses including universities from Romania, Spain, Russia, Slovenia, Portugal and Italy. This study focused on eight factors namely:

1. Learning support: performance of Learning Support Services.
2. Learning delivery: implementation of learning delivery services.
3. Learning development: course design and delivery guidelines, authoring tools and authoring support systems and central support for content developers.
4. Teaching capability: implementation, implemented support services; technical support; staff training for content systems, IT services to support staff.
5. Evaluation: evaluation structures and practices, quality standards and monitoring and review systems.
6. Accessibility: open access policy for students and staff.
7. Technical capability: implementation structure and practice.
8. Institutional capability: institutional ability structures and practices, transfer of research and monitoring results to learning management and practice (Sangrà 2002).

2.5.5 *Sloan's Five Pillars (2002)*

The Sloan Consortium (now known as the Online Learning Consortium) developed five pillars of quality online education. The purpose of Sloan's five pillars was to support technological systems and pedagogical assistance to practices at the educational level. The model of the Sloan Consortium (Sloan-C) is based on quality, good pedagogy and effective practices to improve online education in the learning process. The elements of this model are: learning effectiveness (the curricula and learning resources), access in terms of technological infrastructure (accessibility, reliability, quality of access and technical support services), student satisfaction (course quality, instructor-learner interaction, peer collaboration, and support services), and academic satisfaction (research, professional development and development in Information and Communication Technologies [ICT]) and cost effectiveness (Lorenzo and Moore 2002; Zhao 2003). These five pillars have been used as the basis of a model to evaluate the quality and practice of online learning courses in the universities of the United States for many years (Zhao 2003).

2.5.6 The Policies and Practices of E-Learning in Australia (2005)

In 2004–2005 the Commonwealth Department of Education developed a study to research policies and practices in Australia to provide effective use of technology in the learning process. This study included 400 profiles and policies from 10 developed countries in the world, and six key factors were extracted as necessary conditions for successful e-learning. These factors are as follows: teaching, learning and the curriculum, learning and development, ICT infrastructure, connectivity and Internet, ICT support and innovation (Finger et al. 2006).

2.5.7 Online Learning Success Model (2006)

Holsapple and Lee-Post (2006) provide a framework that forms the basis for their online learning Success Model (see Fig. 2.5). The model includes success metrics developed specifically for the online learning systems being examined. Progress in online learning depends on the realization of achievement and practice at each of the 3 phases of the online process: System design, System delivery, and System outcome (Holsapple and Lee-Post 2006).

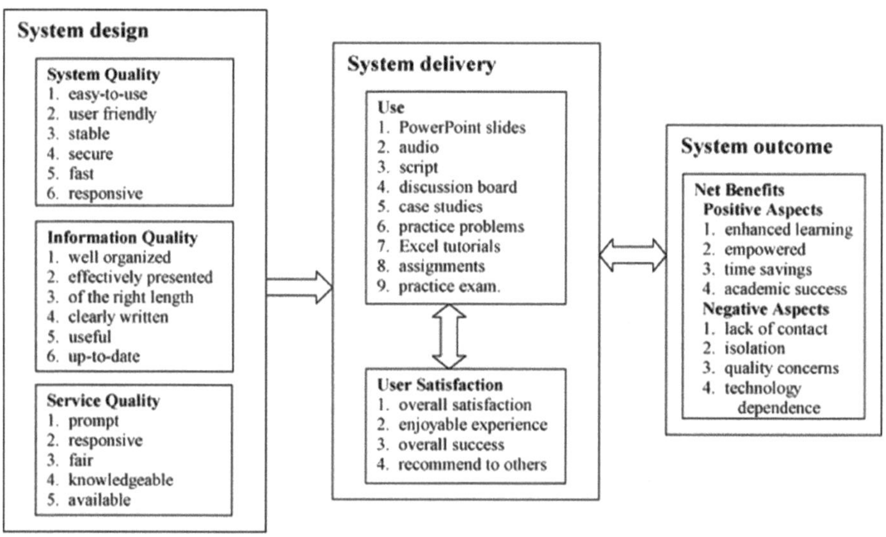

Fig. 2.5 Online learning success model

2.5.8 *The Critical Success Factors in Online Learning (2007)*

Based on Volery and Lord (2000), Selim (2007) focused on critical success factors in online learning and confirmatory factor models. This study specified the online learning success factors through the perspectives of 538 university students. In this study the main factors surveyed and classified into four divisions were:

1. Instructor: IT competency, teaching style and attitude and mindset.
2. Student: time management, discipline, and computer skills and student–student interaction.
3. Information technology: IT infrastructure, reliable and capable of providing the courses and IT tools (network bandwidth, security, accessibility).
4. University support: access to technical advice, technical support and administration support (Selim 2007).

2.5.9 *Khan's Octagonal Framework (2008)*

Khan's Octagonal framework (2008) is one of the main models of integrated e-learning platforms, and has eight dimensions: institutional, pedagogical, technological, interface design, evaluation, management, resource support, and ethical institutional factors (Khan and Granato 2008).

1. Pedagogical: Refers to teaching and learning. This dimension addresses issues concerning content, audiences, goal and media analysis; design approach; organization and methods and strategies of e-learning environments.
2. Technological: Examines issues of technology infrastructure in e-learning environments. This includes infrastructure planning, hardware and software.
3. Interface Design: Refers to the overall look and feel of e-learning programs. The interface design dimension encompasses page and site design, content design, navigation, and usability testing.
4. Evaluation: Includes both assessment of learners, and evaluation of the instruction and learning environment.
5. Management: Refers to the maintenance of learning environment and distribution of information.
6. Resource Support: Examines the online support and resources required to foster meaningful learning environments.
7. Ethical: Relates to social and political influence, cultural diversity, bias, geographical diversity, learner diversity, information accessibility, etiquette, and the legal issues.
8. Institutional: Issues of administrative affairs, academic affairs and student services related to e-learning.

2.5.10 E-Learning Model (2010)

In another contribution, Kala et al. (2010) developed the following E-Learning Model based on constructivism. In relation to educators' roles, this study found three factors of importance for creating an effective teaching–learning environment: (1) enhancing active learning; (2) facilitating social interaction; and (3) creating quality learning materials. In relation to course effectiveness, they found course effectiveness entails assessment of e-learning experiences and involves the use of two types of evaluation processes: (1) quality of the learning materials; and (2) learning outcomes (Kala et al. 2010).

2.5.11 An Analytical Framework to Support E-Learning (2012)

Marshall (2012) developed the following comprehensive model for strategic practice at the level of university in an attempt to create an analytical framework to support e-learning strategy development (Marshall 2012):

- Six critical roles for leaders and managers of learning and teaching
  1. Establishing a vision and direction for the development of learning and teaching;
  2. Aligning stakeholders with this vision and direction;
  3. Motivating and inspiring others to commit themselves to this vision and direction;
  4. Planning and budgeting to support the changes required to realize the vision;
  5. Organizing and staffing to ensure that the work required to effect the change can be efficiently and effectively transacted;
  6. Monitoring and problem solving to ensure that efforts to realize the vision remain "on-track."
- Four specific contexts in which developments need to occur
  1. Program/course/unit;
  2. Faculty/school/department;
  3. Institution; and
  4. Community/sector.
- Four critical domains of practice in which strategies must be developed and implemented
  1. Curriculum development;
  2. Staff development and support;

3. Student learning support; and
4. Institutional enablers (infrastructure) for learning and teaching (organizational, physical and technological).

2.5.12 *Key Success Factors of E-Learning Courses (2012)*

In the study "Key Success Factors of eLearning in Education", FitzPatrick (2012) identified a professional development model to evaluate and support e-learning that has 5 main factors namely Technology (availability, connectivity, and reliability), Human (pedagogy, attitude, and communication), Design (content, interface, and framework), Support (feedback, resources, and training) and Evaluation (assessment, usability, and quality) (FitzPatrick 2012).

2.5.13 *Capability Maturity Model of E-Learning (2012)*

Another attempt to provide a professional model for e learning is the Capability Maturity Model. Zhou (2012) studied key process areas of e-learning. He investigated key success dimensions in a step-by-step process of improvement (FitzPatrick 2012):

1. Learning: pedagogical aspects of e-Learning.
2. Development: creation and maintenance of e-Learning resources.
3. Co-ordination: processes surrounding the oversight and management of e-learning.
4. Evaluation: processes surrounding the evaluation and quality control of e-learning through its entire lifecycle.
5. Organization: processes associated with institutional planning and management.

2.5.14 *A Web-Based Platform for Pedagogy (2013)*

Dragon et al. (2013) developed a web-based platform for pedagogy. This platform rubric was divided into four levels of components (Dragon et al. 2013):

1. High-level component competency: organisation and leadership (effective groups, responsibility for regulating, collective orientation).
2. A second-level component competency: mutual engagement (share deeper concepts and engage in collaborative meaning making).
3. A third-level component competency: support and evaluate the cost and benefit.
4. The fourth-level component competency: feedback on the group dynamics and learning process.

To sum up, in this section an extensive range of literature, largely from higher education settings, was discussed in order to identify a 'baseline' for a framework of practices in e-learning. The e-learning practice researches reviewed originated in different theoretical streams. Some of them were developed based on benchmarking (Barker 1999; Khan 2005; Sangrà et al. 2002; Phipps and Merisotis 2000, etc.), while other e-practice work was based on instructional design theory (Khan 2005) or technology acceptance theory (Volery and Lord 2000; Selim 2007), and so forth. Given this, it can be said that "practice in e-learning" is a variable phenomenon which has been viewed from different perspectives and consequently different aspects have been highlighted, and assorted labels such as Benchmarking, Best Practice, Success Condition etc. have been applied. However, there are large overlaps between the factors of the e-learning practice research despite their different approaches. Indeed, it is hard to target specific factors with discrete approaches because the factors and elements are intermeshed (Oliver 2003).

2.5.15 Summary of Studies Reviewed

Table 2.2 has shown the key factors that are contributing to e-learning practice. The models and guidelines that have been illustrated below are congruent with practical knowledge, and are understandable as best practices based on scientific available evidences. To develop and reach a comprehensive framework for enhancing and assuring practice in e-learning in the present study, a collection of the "previous highlighted evidence" has been reviewed.

2.6 E-Practice Framework

Drawing from the table above, an e-practice framework is constructed on two levels, consisting of "factors" and "sub factors" to underpin best practice. Specifically, the framework is divided into six main factors and 29 sub-factors as listed below:

Pedagogical practice: Student-centred interactivity, socio-communication, learning environment, assessment and learning resources.

Technological practice: Technological infrastructure, functionality of platforms, accessibility, user interface design and reasonably.

Instructional design practices: Clarifying expectations, personalisation, learning scenarios, organizing resources and quality and accuracy.

Organisational practices: Institutional affairs, administrative affairs, research and development (R&D), precedent and reputation.

Support services practices: Administrative support services, technical support services and academic support services.

Table 2.2 Summary of studies reviewed

E-practice	Categories
(American Association of Higher Education) Chickering et al. (1987) Principles for good practice	Encourages contacts between students and faculty Develops reciprocity and cooperation among students Uses active learning techniques Gives prompt feedback Emphasizes time on task Communicates high expectations Respects diverse talents and ways of learning
(Higher Education Policy in the US) Phipps and Merisotis (2000) Benchmarks for success in internet-based education	Institutional support Course development Teaching/learning Course structure Student support Faculty support Evaluation and assessment
(European Commission DG) Sangrà (2002) A benchmark for European virtual campuses	Learning support Learning delivery Learning development Teaching capability Evaluation Accessibility Institutional capability
The model of Sloan Consortium (Sloan-C) (2002) Quality, good pedagogy and effective practices	Learning effectiveness Cost effectiveness Access Academic satisfaction Student satisfaction
(Commonwealth Department of Education) Finger et al. (2006) The policies and practices of e-learning in Australia	Teaching, learning and the curriculum Learning and development ICT infrastructure Connectivity and internet ICT support and innovation
(National Research Council) Putnam and Borko (2000) The learning sciences framework for e-learning systems	Provide contextualized learning Provide scaffolds and tools Students' preconceptions and prior knowledge Organizing conceptual structure explicit Summative assessment of learning Collaborations amongst communities of learners Support learning-by-doing Assessment of learning and knowledge transfer
Jacobson et al. (2005) The pedagogical dimension in the types of e-learning	Role of technology Type of content Role of student Control Learning outcomes

(continued)

Table 2.2 (continued)

E-practice	Categories
Holsapple and Lee-Post (2006) Online learning success model	System design System delivery System outcome
Volery and Lord (2000), Selim (2007) The critical success factors in online learning	Instructor Student interaction Information technology University support
Khan's Octagonal framework (2008) Models of integrated e-learning platforms	Institutional factor Pedagogical factor Technological factor Interface design factor Evaluation factor Management factor Support Ethical institutional factor
Kala et al. (2010) E-learning model	Educator's roles Course effectiveness
Marshall (2012) An analytical framework to support e-learning	Six critical roles for leaders and managers Four specific contexts Four critical domains of practice
FitzPatrick (2012) Key success factors of e-learning in education	Technology Human resource Design Support Evaluation
Zhou (2012) Capability maturity model of e-learning	Pedagogy Development Co-ordination Evaluation Organization
Dragon et al. (2013) A web-based platform for pedagogy	Organization and leadership Mutual engagement Support Feedback

Performance appraisal practices: Cost-effectiveness, learning effectiveness, satisfaction.

Each of these factors and sub-factors, based on the reviewed practice studies and the literature review, will now be briefly described.

2.6.1 Pedagogical Practice Factor

This factor, which addresses the process of learning and teaching in terms of how learning and teaching is done, is at the core of e-learning environments (Masoumi

2010). According to studies reviewed such as Chickering et al. (1987), Sangrà (2002), Finger et al. (2006), Khan (2008) and Zhou (2012), the pedagogical factor is considered to be the most critical in practice. It has five sub-factors.

2.6.1.1 Student Centre Interactivity

Student success can be significantly affected by active engagement in such practices as learning interactivity and integrating past experiences (Chickering and Ehrmann 1996). Interaction and discussion are at the core of a learning and teaching process that can create opportunities for networking and encourage dialogue between and among all the actors in an online learning classroom. According to studies by Chickering and Ehrmann (1996), Phipps and Merisotis (2000), Commissions (2001), Chou (2002) and Zhao (2012), the main focus of student centre interactivity is on student centre practices and activities, interactive networks and discussion in the classroom between and among all the actors.

2.6.1.2 Socio-communication

Socio-communication effectiveness has been defined in a variety of ways (Spitzberg and Hurt 1987; Wiemann 1977). Socio-communicative orientation concerns “one’s approach towards others and how one perceives him/herself, and is much less descriptive of how a person actually behaves” (Frymier 2005, p. 199). Socializing and building a concept of community attracts and retains students in online learning settings (Marshall 2006). A competitive environment, effective communication, facilities and opportunities for good communication and social interactive tools are counted as influential factors in the success of e-learning (cf. Reeves and Reeves 1997; Oliver and Herrington 2007; Frymier 2005; Zhao 2012).

2.6.1.3 Learning Environment

Environmental learning facilities allude to locations, contexts, settings and cultures in which students learn (Fraser and Fisher 1994). Students who keenly support the collaborative wiki tool are successful at using it to complete unit tasks in a flexible online environment (Raitman et al. 2005). Creating and improving a sense of space and feeling at home could be important elements in reducing the dropout rate between online students (Masoumi 2010). A flexible environment system and environmental learning facilities are the main factors of an effective learning environment (cf. Achtemeier and Simpson 2005; Raitman et al. 2005).

2.6.1.4 Assessment

Assessment can focus on the students' progress, the learning community, teacher practices, e-learning systems and organisation. The assessment of learning in online programs requires policies, practice and tools that are clear, valid, reliable, and can be automatically administered and scored (cf. Thompson et al. 2015). "Assessment in e-learning can be carried out in different modes by teachers, peers, by means of self-assessment as well as the students' portfolios" (Masoumi 2010, p. 196). Assessment in online programs, though, can be challenging due to problems of classroom feedback, academic honesty, plagiarism and feedback on assessment results (cf. Gáti and Kártyás 2011; Kala et al. 2010; McKinnon et al. 2000; Phipps and Merisotis 2000; Wahlstedt et al. 2008).

2.6.1.5 Learning Resources

Having adequate contents is a necessary priority in academic setting. Hostager's (2014) research findings show that adequate learning resources and services have a positive effect on the grades that students earn in e-learning programs (Hostager 2014). E-learning providers are expected to provide a variety of e-resources to support learners' learning practices and activities (cf. Finger et al. 2006; FitzPatrick 2012; Kala et al. 2010; Khan and Granato 2008; Marshall 2012; Phipps and Merisotis 2000).

2.6.2 Instructional Design Practices (Curriculum) Factor

According to Gagne et al. (2005) and Laurillard (2013), this factor is "an iterative process that refers to the structuring and arranging of resources and procedures used to promote learning in an institution" (Masoumi 2010, p. 190). Instructional design issue is "the first important one related to usability and efficiency of a user interface" (Skalka et al. 2012, p. 3). One of the best e-practices is related to the instructional design issue which concerns the framing of all the elements of the learning process in order to create effective learning and teaching environments (cf. Phipps and Merisotis 2000; Finger et al. 2006; Putnam and Borko 2000; Marshall 2012). It has five sub-factors.

2.6.2.1 Clear Expectations

Learning audiences tend to focus more on learning when e-learning programs are organized with clear expectations (Ku et al. 2011). Clear explanation is a key to successful e-learning programs because clear explanation helps to prevent misunderstanding of content of learning and tasks (Lee 2014). Clear objectives, expectations and syllabi prefigure unity between learning activities by describing the learning content, the actions to be taken or performed and how these will be assessed (cf. Phipps and Merisotis 2000; Holsapple and Lee-Post 2006; Kala et al. 2010; Khan and Granato 2008; Marshall 2012; Lee 2014)

2.6.2.2 Learning Scenarios

Present scenarios of e-learning programs are taking advantage of online web technologies to connect learners and facilitate sharing information in an interoperable way for satisfactory learning experience based on effective scenarios (Santos and Boticario 2015). As Marshall (2006) explained, the online scenario, which can be considered to be an educational technique, can shape and influence every part of the learning process, both as a means of understanding how students learn and as tools for guiding the design and aligning of learning activities and practices (Masoumi 2010). The scenarios of online learning should be selected based on the goals of the course, content of modules and effective instructional strategies (cf. Chickering et al. 1987; Duffy et al. 2012; FitzPatrick 2012; Kala et al. 2010; Oliver 2001).

2.6.2.3 Accuracy of Resources

The accuracy of resources sub factor is an important one which is related to the reliability of the instructional materials in e-learning (cf. Phipps and Merisotis 2000; Holsapple and Lee-Post 2006; Zhao 2012).

2.6.2.4 Organizing Resources

Organizing resources incorporates different activities and practices, e.g. sorting or grouping resources of interest in a personal classification system, storing of organized content, at least for the time of use and sharing of the arranged content with peers (Seidel 2014, p. 6). According to Oliver (2001), Holsapple and Lee-Post (2006) and Lee and Rozinah (2009), the main quality issues concerning organizing and structuring learning resources can ultimately determine the effectiveness and efficiency of the learning environment (Masoumi 2010).

2.6.2.5 Virtual Personalization

In order to improve success in e-learning practice, additional interventions in online programs need to be explored, including those that increase student motivation via personalization (Pemberton and Moallem 2013). As Martinez (2010) explained, "personalization uses student-specific approaches to address individual needs and expectations to support and promote individual learning success" (Pemberton and Moallem 2013, p. 908). In fact students' motivation increases as a result of a personalized link between the students and the content, and is directly affected by the manner in which the content is presented to the students (Wlodkowski 1999). Virtual environments based on students' needs and interests directly affect the learning and teaching process (Klašnja-Milićević et al. 2011; Marshall 2012; Weld et al. 2012; Pemberton and Moallem 2013).

2.6.3 *Technological Practice Factor*

Studying technological factors and infrastructures in e-learning practices will involve paying attention to the interconnectedness of artifacts and tools and seeing the different tools as part of and embedded in social, institutional, and technological arrangements (Guribye 2015). The IT infrastructure is viewed as a 'web' of equipment, or app whose efficiency can be determined in terms of availability and reliability, providing adequate functionalities, integration into the existing infrastructure (Guribye 2005). Based on a large number of studies in this area (cf. Finger et al. 2006; FitzPatrick 2012; Guribye 2015; Holsapple and Lee-Post 2006; Jacobson et al. 2005; Khan 2008; Sangrà 2002; Selim 2007; Volery and Lord 2000), it has five sub-factors.

2.6.3.1 Technological Development

An important aspect of infrastructure development for learning purposes is attention to how social and pedagogical arrangements are related to the tools and technologies of a given setting (Guribye 2015). Technological development includes infrastructural development and IT development (cf. Dragon et al. 2013; Finger et al. 2006; FitzPatrick 2012; Holsapple and Lee-Post 2006; Khan and Granato 2008; Lorenzo and Moore 2002; Phipps and Merisotis 2000; Selim 2007).

2.6.3.2 Functionality of Platforms

The functionality of a learning management system is key to administrating training activities and delivering online training (Guribye 2015). The functionality of infrastructure includes e-learning platforms and LMS capabilities (cf. Wirth 2005; Guribye 2015).

2.6.3.3 Accessibility

"Broadly speaking, accessibility usually goes beyond computers and connections" (Masumi 2010, p. 188). Accessibility facilitates the delivery of learning materials and is concerned particularly to include those students who may not have equal access through some form of disability. The e-accessibility can, with sufficient planning and system wide approaches being adopted, help promote the inclusion of learners with various disabilities and problems (Di Iorio et al. 2006). There has been research related to the e-accessibility needs and problems of learners with disabilities, including online programs, where the students were supported by campus disability service providers and other support services (Stodden et al. 2006). These included assistive technologists and disability officers aiding the e-learning experiences of students (Thompson 2004).

Flexibility, unlimited and easy access to technology and networks are the main key items in the accessiblity subfactor area. (Commissions 2001; Fichten et al. 2009; Lau et al. 2014; Sangrà 2002; Selim 2007).

2.6.3.4 Reasonable

Reasonable includes sharing educational material, ensuring future development, guidelines, strategies of information technology (Díez et al. 2012; Lavrischeva and Ostrovski 2012; Marshall 2012; Selim 2007).

2.6.3.5 User Interface Design

User interface design includes platforms based on a user-friendly environment, platforms based on self-evidence and predictability, platforms based on learning activities going smoothly and effectively, based on highly standardized navigation (cf. Guo et al. 2010; Guo et al. 2009; Holsapple and Lee-Post 2006; Rosenbaum 2013).

This way of approaching infrastructures as a set of resources in networked form clearly builds on and adds to the picture of infrastructure conveyed through the studies just cited. Although there are many similarities between knowledge work and educational activities or learning practices, there are also some key differences in the focus and the object of the activities. Although I argue that we can borrow from the perspective of infrastructure as relational and ecological and bring the same sensitivities to the study of networked learning and technology-enhanced learning, there are some inherent concerns within these fields that must be considered, such as the notion of a pedagogical approach (Guribye 2015).

2.6.4 Organisational Practice Factor

Within the higher education system, e-learning becomes an organisational practice focused on institutional policy and processes (De Freitas and Oliver 2005). As Marshall (2006) discussed, a successful online program depends on explicit organisational strategies and aims, along with well-established procedures and effective rules and standards (Masoumi 2010). Institutional affairs, administrative affairs, research and development and precedent and reputation are the four main sub factors related to organisational practice (cf. Dragon et al. 2013; Khan 2008; Novak 2002; Sangrà 2002).

2.6.4.1 Institutional Affairs

Institutional affairs and clear and effective practice are focused on structures and practical procedures such as students' affairs and online programs (Oliver 2003).

2.6.4.2 Precedent and Reputation

University e-learning reputation is an important element related to organizational success or failure. In fact as Moore (2005) discussed, the successful implementation of programs and organisational policies is one of the clear ways to achieve precedent and reputation.

2.6.4.3 Administrative Affairs

"The institutional dimension of e-learning is concerned with issues of administrative affairs related to e-learning" (Khan 2005, p. 25). A large number of studies such as Frydenberg (2002), Inglis (2005) and Wirth (2005) showed that supportive administrative affairs have a strong effect on establishing and maintaining quality in learning systems.

2.6.4.4 Research and Development (R&D)

As Dirr (2003) and Kyvik (2009) discuss, the interdependence of R&D is the foundation of educational success (Masoumi 2010). In fact the sub-factor of R&D focuses on research opportunities and facilities based on strategies and goals in the e-learning program system.

2.6.5 *Support Services Practice Factor*

A help desk support service is an important practical strategy for more effective online programs and also it contributes to student university success (Karabenick and Newman 2013; Schworm and Gruber 2012). Moore and Fetzner (2009) synthesized a variety of student support factors that have contributed to high course completion rates, including personalized access to administrative and programmatic contacts; advisors and coaches; online and/or on-campus orientations to online learning; a 24/7 technical support help desk; academic support and tutoring; and enabling students to support each other through online community websites, courses or student associations (Milman, Posey, Wright and Zhou 2015, p. 7). In fact both learners and teachers need an academic or technical service and successful support in e-learning programs (Schworm and Gruber 2012). Academic or pedagogical, administrative and technical support services are the three main kinds of support, especially in the learning and teaching process and problem solving (cf. Finger et al. 2006; FitzPatrick 2012; Khan 2008; Phipps and Merisotis 2000; Sangrà 2002; Selim 2007; Volery and Lord 2000).

2.6.5.1 Technical Support

An IT support service is an integral part of any successful online program for all learners, lecturers and providers (Masoumi 2010). In fact for lecturer and student-related technical problems, help desk, fast feedback and technical problem solving are the main issues in this sub factor (cf. Dragon et al. 2013; Finger et al. 2006; FitzPatrick 2012; Khan and Granato 2008; Lorenzo and Moore 2002; Marshall 2012; Phipps and Merisotis 2000; Sangrà 2002; Selim 2007).

2.6.5.2 Administrative Support

An understanding of the e-learning program learners' perspective related to administrative support was identified in the studies as important to student performance and retention (Milman et al. 2015). In fact the university system needs administrative services (such as financial, supplies, control, accountability system and resources) supporting all enquiries based on student priority (cf. Calvo 2007; Milman et al. 2015; Sims et al. 2002).

2.6.5.3 Academic Support Services

The students of e-learning programs require academic and tutoring assistance toward effective and successful learning (Milman et al. 2015). Academic support services with a special emphasis on e-Learning pedagogical and professional procedures are needed in the learning and teaching process rather than simply technological support services (Marshall 2006). According to studies like Ellis and Calvo (2007) and Milman et al. (2015) it is necessary for a successful learning and teaching process in an online environment.

2.6.6 Performance Appraisal Practice Factor

A performance appraisal e-learning practice is a core part of the educational setting and system of a university. Performance appraisal practice, "as a meta-indicator, is used to stress the ability of an institution to produce the desired result as measuring criteria for how, and the extent to which, it meets the demands at different levels" (Masoumi and Lindström 2012, p. 34). Learning effectiveness, satisfaction and cost-effectiveness are the main sub factors related to e-learning evaluation and performance appraisal (cf. FitzPatrick 2012; Kala et al. 2010; Khan 2005; Merisotis 2000; Phipps and Sangrà 2002; Sloan-C 2002 and Zhou 2012).

2.6.6.1 Cost-Effectiveness

The strategy of e-learning is considered a cost-effective way to allow distant learners to have access to online programs (Aguti et al. 2014). Along with development of online learning programs, the analysis of cost-effectiveness based on budgetary constraints, limited time frame, logistical boundaries and providing effective and satisfactory proportions is necessary and a main priority (cf. Masoumi and Lindström 2012; Moore 2005 and Sloan-C 2002).

2.6.6.2 Learning Effectiveness

One of the important concerns of e-learning providers is effectiveness of performance and outcome of learning programs in the online environment (Masoumi and Lindström 2012). Learning effectiveness studies have focussed on different ways to assess the effectiveness of any intervention in the learning and teaching process such as learners' features, prior experience, and interactive discussions and feedback on performance, standards of quality of outcomes (cf. Aguti et al. 2014; Khan 2010; Marshall 2006).

2.6.6.3 Satisfaction

The results of several studies showed that satisfaction is an important and influential component in the quality of e-learning programs (cf. Chen et al. 2011; Marki et al. 2000; Wang 2003). Achieving satisfactory performance, motivation to achieve outcomes and e-learning experience satisfaction are the main effective elements related to students and lecturer performance improvement in online programs (cf. Al-Huneidi and Schreurs 2011; Alzahrani and Ghinea 2012; Chickering et al. 1987; Holsapple and Lee-Post 2006; Juwah et al. 2000, 2013; Marshall 2012; Zhao 2003).

This chapter was organized into three parts. An overview of the literature dealing with historical background, e-learning definition, cultural context in e-learning and theoretical and practical research studies were provided.

The following chapters present the results of the qualitative and quantitative studies undertaken in this study. One of those is an Australian university and the other is American. The results of the two countries studied are then compared. This comparative study adopted a mixed method case study research design. The current status and the dominant cultural dimensions of e-learning practice were investigated in the two universities using a predominantly quantitative methodological approach. The four major aspects of e-learning practice, namely pedagogy, culture, technology and e-practice, were investigated by applying a qualitative method.

The qualitative approach undertaken in this study predominantly involved interviews by the researcher. "As noted by Creswell and Plano Clarke (2007), qualitative data is made up of open-ended information primarily gathered from participant interviews" (Bagnall 2015, p. 29). According to the methodological assumptions underpinning a qualitative study these include: using inductive logic, researching the topic within its context and using an emerging design (Creswell 2013). The methodological approach to the study of the current state of e-learning practice is firmly based within this comparative context. The main strategy of the comparative framework "... has to progress from accurate description to analysis, and from that to the forming of generalizations about the working of educational systems"(Grant 1999, p. 66). In addition to adopting a mixed methodology, as mentioned earlier, the researcher considered a case study approach as an appropriate research design for conducting this study. The purpose of a case study is to provide an analysis of a contemporary phenomenon to gain a robust description and analysis of a single case or unit (Merriam 1988). "Studies of a phenomenon as subtle and complex as the learning of science require in-depth examination of individual learners" (Taber 2000, p. 469). However, it is often considered that analytics researches are not generalizable as the results can only claim any credibility in the setting of the particular cases researched (Walker 1980), although it is recognized that they may well provide insights of wider value. However, "... other authors expand the definition of generalizability beyond its traditional normative meaning ('statistical generalization'), to include 'analytical generalization" (Taber 2000, p. 470). This "involves a reasoned judgement about the extent to which the findings from one study can be used as a guide to what might occur in another situation" (Kvale 1996, p. 233).

The adoption of such a methodology means that generalizability from the case studies may be problematic. It may be argued that generalizability can be increased by the strategic selection of cases studies. (Flyvbjerg 2006). Merriam (1988) noted the true value of a case study is the intimate relationships that exist within that unit that the researcher has access to (Merriam 1988; Stake 1995). Yin (2014) noted the challenges of case studies are rigor and generalizability. Mixing qualitative and quantitative research adds to the richness of case study research (Creswell 2013). Investigating an Australian university and an American university and their e-learning practices through a qualitative method, constitutes an urgent endeavor because quantitative methods only provide superficial feedback. Therefore, it is important that the superficial findings generated from quantitative research about both countries online learning programs be complemented with qualitative research, which can provide more complexity and a richer picture of the current e-practice issues under investigation. The next chapter of the book presents the dominant cultural orientations of e-learning in one Australian university and one American university.

Chapter 3
The Dominant Cultural Dimensions in Comparative Context

3.1 Introduction

E-learning, as argued, is influenced by culture and thus embedded in a cultural context. Correspondingly, it is impossible to decontextualize and separate these initiatives from their context and the circumstance in which they are going to be used. In a same vein, teaching and learning are embedded in the cultural context and, as Henderson (1996) put forward, cannot and do not exist outside of cultural contexts. Subsequently, cultural dimensions are an integral part of every aspect of the educational system including e-learning (Edmundson 2003).

Ehlers (2009) provided a culturally-sensitive model for enhancing quality in online learning programs (Masoumi 2010). The claim is that e-learning programs should be relevant to the context of the culture in which they have been applied. To achieve greater success from e-learning programs developed countries like the United States of America and Australia attempt to investigate how to individualise characteristics, technology and contexts of their e-learning system (Anderson and Gronlund 2009).

Cultural context and cultural dimensions are essential aspects of e-learning systems that both directly and indirectly affect their quality (Edmundson 2003; Masoumi 2010). Therefore cultural factors can be seen as the foundation for furnishing improved e-learning systems that can modify the whole e-learning structure (Kujala and Lillrank 2004). Cultural aspects such as educational paradigms, origin of motivation, experimental values, value of errors, role of instructor, user activity, learner control, accommodation of individual differences and collaborative learning (see Edmondson 2004; Gamble 2009; Henderson 1996; Masoumi and Lindström 2012; Reeves and Harmon 1994; Washburn 2012) focus on the pedagogical context of e-practice (Reeves and Reeves 1997) which may be oriented towards either constructivism or instructivism.

A large number of studies (cf. Butter et al. 2015; Chen et al. 2012; Collis 1999; Edmundson 2004; Henderson 1996; Reeves 1994; Usun 2004; Masoumi and

S.H. Sadeghi, *E-Learning Practice in Higher Education: A Mixed-Method Comparative Analysis*, Studies in Systems, Decision and Control 122,
https://doi.org/10.1007/978-3-319-65939-8_3

Lindström 2012) have investigated cultural issues in online learning. In fact many researchers, such as Reeves (1994), Henderson (1996) and McLaughlin (2000) have focused specifically on designing models for assessing the cultural dimensions of education and educational artefacts, such as e-learning.

The first aim of this current research is to compare the dominant cultural dimensions between one faculty in an Australian university and one faculty in a US university in order to ultimately improve the quality of online learning courses in Australian and American universities. Accordingly, Edmundson's questionnaire addressing the cultural dimensions was adopted. To provide a comprehensive understanding of the cultural dimensions, these dimensions are summarized in Table 3.1.

The results of this study are reported by considering these factors based on student, lecturer and administrative staff perspectives (see "Method" section for more information). Each factor based on these variables has been analysed separately in each country and then the results of comparing the countries on that factor are reported to identify similarities and differences in their dominant cultural dimensions.

The last part of this study is the discussion. In the discussion the final conclusion regarding the results, their means as well as a short summary of other research that has been conducted previously are reported. Because the researcher could not locate

Table 3.1 Cultural dimensions

Dimensions	Instructivism	Constructivism
Educational paradigm	Behavioural approach predetermined goals based on knowledge acquisition	Constructivist approaches unfocused goals based on knowledge transfer
Experimental values	Learning practices based on abstract approaches	Learning practices based on real world
Role of instructor	Lecturer-centred teaching	student-centred teaching
Value of errors	Fulfill a instruction course without making mistakes	Mistakes as part of the learning process
Origin of motivation	External interest and needs	Internal motivation and true desire
Accommodation of individual differences	Single-faceted consideration on learners affective and physiological differences accommodated in learning environments	Multi-faceted consideration on learners' needs and preferences based on affective and physiological differences
Learner control	Students learning program is predetermined and fully controlled	Students have power to choose what section, and/or what paths to follow
User activity	Students access various representations of content limited in predetermined path	Students engage in the learning process for creating and managing knowledge as main user
Collaborative learning	Limited support and no facilities for setting up collaborating learning	Variety of different facilities and support are provided for setting up collaborative learning

any research comparing Australia and America based on their cultural dimensions, the results of comparing other countries and America have been summarised for each factor drawn from the work of Edmonson (2004) who has focused on comparing India and America, Gamble (2009) who compared China and America, Morris (2009) and Washburn (2012) who analysed Asian and American e-learning systems. At the end of the discussion some limitations of the current study are described and some suggestions for further research are provided.

3.2 Method

3.2.1 Participants

205 participants from one faculty in an Australian university (n = 99) and one faculty in an American University (n = 106) were recruited to participate in this study through an online invitation email requesting volunteers. From Australia, 46 participants were female, and 53 were male. They reported their age as 20–30 (n = 34), 30–40 (n = 44) and 40–50 years (n = 21). Seventy participants were students, 18 were lecturers and 11 were administrative staff. They reported their online experiences based on two categories namely beginner to average (people who have participated in 1–4 courses; n = 35), and average to expert (people who have participated in up to 4 courses; n = 64). From United States of America, 61 participants were female, and 45 were male. They reported their age as 20–30 (n = 35), 30–40 (n = 41) and 40–50 years (n = 30). Seventy two participants were students, 21 were lecturers and 13 were administrative staff. Participants reported their online experiences as beginner to average (n = 64), and average to expert (n = 42). Normality of distributions and homogeneity of variances were checked. The results indicated no outlier. Table 3.2 is a summary of the demographic makeup of the participants of both countries.

Table 3.2 Demographic information based on country

Country	Gender	*N*	Age	*N*	Position	*N*	Experience	*N*
AUS	Female	46	20–30	34	Student	70	Beginner to average	35
	Male	53	30–40	44	Lecturers	18	Average to expert	64
	–	–	40–50	21	Staff	11	–	–
	Total	99						
USA	Female	61	20–30	35	Student	72	Beginner to average	64
	Male	45	30–40	41	Lecturers	21	Average to expert	42
	–	–	40–50	30	Staff	13	–	–
	Total	106						
All total						205		

3.2.2 Measures

To explore the comparative evaluation of e-learning e-practice factors in an Australian university and a USA university, a researcher-constructed survey was applied consisting of 18 questions measuring 6 cultural pedagogical factors namely educational paradigm, collaborative learning, value of errors and accommodation of individual differences and role of instructor and learner control. Each factor was measured by two or three questions with two statements which examined different dimensions of instructivism or constructivism. Reliability, content validity and construct validity of this questionnaire have been reported by Edmundson (2004).

Because this questionnaire did not cover different dimensions of cultural-pedagogical construct, three more factors from Masumi (2012) were added to the Edmundson questionnaire which have been included in this research study.

Further, three more factors namely experimental values, user activity and origin of motivation have been taken from Reeves (1994) and Henderson (1996) and added to the main questionnaire. Similar to the Edmundson questionnaire, these factors have been represented by two or three items with two statements to measure the dimensions of instructivism and constructivism, so the final questionnaire included 21 items in total.

It is to be noted that questions 1, 2 and 3 measured the factor of educational paradigm, questions 4, 5 and 6 measured the factor of experimental value, questions 7 and 8 measured the factor of the role of instructor, questions 9 and 10 measured the factor of value of errors, questions 11, 12 and 13 measured the factor of origin of motivation, questions 14 and 15 measured accommodation of individual differences, questions 16 and 17 measured the factor of learner control, questions 18 and 19 measured the factor of user activity and finally questions 20 and 21 measured the factor of collaborative learning. The researcher modified two versions of this questionnaire, one for students and one for lecturers or administrative staff (see Appendix B).

3.2.3 Design and Procedure

The researcher first started to investigate Australian and American universities offering e-learning courses according to rankings of the universities and their program details in order to find the universities with the best e-learning courses.

Because of ethical restrictions the researcher could not write the name of either university. It has been stressed through out the book that this study is based on a small sample of only two Universities, one Faculty in an Australian university and one Faculty in an American university.

For Australia, the researcher sent several official letters to conduct this study in Australian universities like Curtin University, UNSW, Western Sydney University and Queensland Technology University. However, the directors did not agree to cooperate with the researcher. Ultimately, the University of X was selected due to the fact that the research was conducted in that university, plus accessibility to a variety of facilities and the ability to negotiate with different schools that provided e-learning courses.

The University of Y was selected as the sample of American university. Several American universities including Pennsylvania State University, Boston University and University of Florida were approached but the University of Y was the one that accepted. After selecting X and Y Universities, the researcher started to investigate different online courses in the faculties. In the University of X of the faculties of business, IT, engineering and health sciences, just the faculty of health sciences agreed to participate in this study, so the researcher decided to select the school of public health in Y University as the alignment. Similarity of the subject of study was the main reason for this selection. Finally the participants were chosen based on their availability and willingness to participate. Ethical approval was obtained to conduct this study in both universities and the survey was made by applying Lime Survey software.

Thereafter, the link of the survey was sent by email to students, lecturers and administrative staff in each university.

In the first section of the survey, demographic information regarding gender, age of participants in three levels 20–30, 30–40 and 40–50 years and academic position in three categories of student, lecturer and administrative staff was sought. Also they were asked to indicate their experience in engaging with the e-learning system based on two categories, beginner to average (people who have 1–4 online courses), and average to expert (people who have up to 4 online courses). The second part of the survey was the questionnaire. The last part was a thank you page. After collecting data, SPSS program was applied for analysing the data. The researcher coded each response to each question of the questionnaire in which 1 represented the instructivism orientation and 0 represented the constructivism orientation of each dimension. A descriptive analysis including frequencies and percentage of each question was conducted for each country based on academic position. To compare the results of Australia and America, a chi-square test was applied which has been reported in the result section.

3.3 Results and Key Findings

This section is concerned with the cultural aspects of the two universities, one faculty in an Australian university and one faculty in a USA university. The two dimensions of instructivism and constructivism are dealt with in relation to these cultural dimensions. Firstly the results of each question in relation to academic position in the Australian sample are discussed. Secondly the results of each individual factor in relation to these variables within the American sample are dealt with. Subsequently comparison of the results of each individual factor within the American and Australian sample is discussed.

3.3.1 *Educational Paradigm*

As explained in the method section, the educational paradigm was measured by 3 main questions with two orientations, sharply focused based on knowledge acquisition and unfocused based on knowledge transfer. The participants were requested to choose one of the orientations based on their understanding of the e-learning program which they had engaged with in their educational system. The orientation of sharply focused (knowledge acquisition) indicates the instructivism aspect and the orientation of unfocused (knowledge transfer) indicates the constructivism aspect of the educational paradigm dimension.

It is worth mentioning that the first question is about the path to learn that students apply to their e-learning system, the second question is about the approach to assessment of students who have been assessed on this system and the third question is about students' learning goals. To contribute a broader understanding of the similarities and differences between the educational paradigm dimensions of Australian and American participants, comparative results are reported based on academic position.

3.3.1.1 Question 1: Path to Learn

Table 3.3 illustrates the frequency, and percentage of responses to question 1 in participants of one faculty in an Australian university and participants of one faculty in an American university based on their academic position (students, lecturers and administrative staff). This first question evaluates the path to learn of students using either the orientation of instructivism or constructivism.

As can be seen in this Table, the results showed that in an Australian university, 61.4% of students (n = 43), 44.4% of lecturers (n = 8) and 27.3% (n = 3) of administrative staff believed in a sharply focused path to learn of students based on Knowledge acquisition. However, 38.6% of students (n = 27), 55.6% (n = 10) lecturers and 72.7% of administrative staff (n = 8) believed in an unfocused path to learn of students—based on knowledge transfer—regarding their e-learning environment. According to these results, it seems that the dominant aspect of Australian

Table 3.3 Comparison of responses to path to learn

Path to learn	Paradigm	Students		Lecturers		Staff		χ^2
		N	%	N	%	N	%	
AUS participants	Sharply focused	43	61.4	8	44.4	3	27.3	5.37
	Unfocused	27	38.6	10	55.6	8	72.7	
USA participants	Sharply focused	42	58.3	5	23.8	3	23.1	11.22*
	Unfocused	30	41.7	16	76.2	10	76.9	

Sharply focused (Knowledge acquisition) = Instructivisim; Unfocused (Knowledge transfer) = Constructivism
$^*p < 0.05$

students' path to learn is instructivism; however, the dominant aspect of Australian lecturers and administrative staff is constructivism. However, based on the Chi-square test, there were no significant differences between students, lecturers and administrative staff's perspectives on this question [χ^2 (2, $N = 99$) = 5.37, $p = 0.06$].

In an American university, as reported in Table 3.2, 58.3% of students ($n = 42$), 23.8% of lecturers ($n = 5$) and 23.1% ($n = 3$) of administrative staff believed in a sharply focused path to learn of students based on Knowledge acquisition in their e-learning system. However, 41.7% of students ($n = 30$), 76.2% ($n = 16$) lecturers and 76.9% of administrative staff ($n = 10$) believed that students try to create and explore their own path to learn unfocused based on their e-learning environment. From this perspective, more than half of the students believed in instructivism in the path to learn for students, whereas the majority of lecturers and administrative staff believed in a constructivism path to learn for students. Based on the Chi-square test, there were significant differences between students, lecturers and administrative staff's perspectives in America on this question [χ^2 (2, $N = 106$) = 11.22, $p = 0.004$]. This difference between American students on one hand and American lecturers or administrative staff on the other hand about choosing a path to learn can be explained by comparing the academic positions of participants. It seems that ideally, lecturers and administrative staff in America prefer that students design and follow their own path to learn, however, from the perspective of students, they tend to follow a well-defined, logical path to learn that has been defined to them by the system. This tendency in students may be due to their lack of knowledge, their skills or abilities to define their own path to learn. Also miscommunications between lecturers or administrative staff and students about the importance of defining a self-fitted path to learn may cause this difference in their perspectives. Teaching how to define your own path to learn and how important it is to students may fill this gap between the perspective of lecturers or administrative staff and students.

The results of dominant aspects of each country based on their academic positions about path to learn show that there were no significant differences in the dominant aspect of students about path to learn in both countries [χ^2 (1, $N = 142$) = 0.14, $p = 0.70$]. The dominant aspect of students about path to learn in both countries was sharply focused and toward instructivism orientation. This means that students in both countries tend to follow the well-defined logical path to learn that has been defined to them by the educational system and not by themselves.

To continue, the results of the Chi-square test showed that there were no significant differences in the dominant aspect of Australian and American lecturers about the path to learn of students [χ^2 (1, $N = 39$) = 1.85, $p = .17$]. This result showed that there is similarity in the perspectives of Australian and American lecturers about the path to learn of their students which is toward constructivism. In both countries lecturers tend to believe that their students define their own path to learn, a path which suits them.

Moreover, the results of the Chi-square test revealed that there were no significant differences in the dominant aspect of Australian and American administrative staff about question 1 [χ^2 (1, $N = 24$) = 0.05, $p = 0.81$]. The dominant perspective of Australian administrative staff about the path to learn of students is toward instructivism; however, the dominant aspect of American administrative staff about the path to learn of their students is toward constructivism. This difference may be due to the fact that the e-learning system in America encourages more administrative staff to let students find, follow and create their own path to learn.

3.3.1.2 Question 2: Approach of Assessment

The second question of the educational paradigm is about the approach an e-learning educational system uses to assess students' performance. As has been indicated in Table 3.4, 65.7% of Australian students (n = 46) and 72.2% of lecturers (n = 13) and 54.5% of administrative staff (n = 5) explained that students are usually tested with questions that are based on the stated goals and objectives of the course for knowledge acquisition which shows a sharply focused orientation of participants to this question. However, 34.3% of students (n = 24) and 27.8% of lecturers (n = 5) and 45.5% of administrative staff (n = 6) explained that students are tested by applying what they have learned from the course to different situations. According to these results, the majority of students, lecturers and administrative staff believed in a sharply focused approach to assessment in their educational e-learning system which shows that the dominant aspect of all participants is toward instructivism [χ^2 (2, N = 99) = 2.26, p = 0.32].

Considering this question in an American University showed that 41.7% of students (n = 30) and 42.9% of lecturers (n = 9) and 46.2% of administrative staff (n = 6) indicated their approach to assessment had a sharply focused orientation toward knowledge acquisition. However, 58.3% of students (n = 42) and 57.1% of lecturers (n = 12) 53.8% of administrative staff (n = 7) were oriented toward an unfocused approach to assessment with a knowledge transfer perspective. According to these results, more than half of the students, lecturers and administrative staff believed in an unfocused approach to assessment in their educational e-learning system which indicated that the dominant aspect of all the Americans to this question is toward a constructivism orientation [χ^2 (2, N = 106) = 0.09, p = 0.95].

Table 3.4 Comparison of responses to approach of assessment

Approach of assessment	Paradigm	Students		Lecturers		Staff		χ^2
		N	%	N	%	N	%	
AUS participants	Sharply focused	46	65.7	13	72.2	5	54.5	2.26
	Unfocused	24	34.3	5	27.8	6	45.5	
USA participants	Sharply focused	30	41.7	9	42.9	6	46.2	0.09
	Unfocused	42	58.3	12	57.1	7	53.8	

Sharply focused (Knowledge acquisition) = Instructivisim; Unfocused (Knowledge transfer) = Constructivism

Comparing the results of the second question shows that there are significant differences between the dominant aspect of students' perspectives in an Australian university and a American university [χ^2 (1, $N = 142$) = 8.25, $p = 0.004$]. The dominant aspect of Australian students about their approach to assessment is toward instructivism, however, the dominant aspect of American students is toward constructivism. The results of Chi-square test showed that there were no significant differences in the dominant aspect of Australian and American lecturers about this question [χ^2 (1, $N = 39$) = 3.39, $p = 0.06$]. In addition, the results of Chi-square test showed there were no significant differences between Australian and American administrative staff' perspectives on this question [χ^2 (1, $N = 24$) = 0.001, $p = 0.97$].

3.3.1.3 Question 3: Learning Goals

The third question regarding the educational paradigm dimension attempted to measure the orientation of learning goals in students in e-learning systems. As can be seen in Table 3.5 in Australia 70% of students ($n = 49$) and 66.7% of the lecturers ($n = 12$) and 54.5% of administrative staff ($n = 6$) assessed sharply focused for learning goals of students claiming that students are given predetermined learning goals. However, 30% of the students ($n = 21$), 33.3% of the lecturers ($n = 6$) and 45.5% of administrative staff ($n = 5$) explained that students learn as they go, depending on their own learning goals. These results illustrated that most students, lecturers and administrative staff in Australia tended to have an instructivism orientation in relation to learning goals [χ^2 (2, $N = 99$) = 1.04, $p = 0.59$].

Considering the third question in America about learning goals revealed that 65.3% of American students ($n = 47$) and 57.1% of the lecturers ($n = 12$) and 61.5% of administrative staff ($n = 8$) assessed sharply focused for learning goals of students based on knowledge acquisition. However, 34.7% of the students ($n = 25$), 42.9% of the lecturers ($n = 9$) and 38.5% of administrative staff ($n = 5$) assessed learning goals as having unfocused orientation based on knowledge transfer. These results illustrated that most students, and administrative staff tended to have an instructivism orientation in relation to their learning goals. However, more than half of the lecturers believed in a constructivism orientation of their students' learning

Table 3.5 Comparison of responses to learning goals

Approach of assessment	Paradigm	Students		Lecturers		Staff		χ^2
		N	%	N	%	N	%	
AUS participants	Sharply focused	49	70	12	66.7	6	54.5	1.04
	Unfocused	21	30	6	33.3	5	45.5	
USA participants	Sharply focused	47	65.3	12	57.1	8	61.5	0.48
	Unfocused	25	34.7	9	42.9	5	38.5	

Sharply focused (Knowledge acquisition) = Instructivisim; Unfocused (Knowledge transfer) = Constructivism goals. However, there were no significant differences between the dominant aspect of students, or administrative staff and lecturers to this question [χ^2 (2, $N = 106$) = 0.48, $p = 0.78$].

Comparing the results of Australians and Americans about the third question showed there were no significant differences between the dominant aspect of Australian and American students' perspectives on this question which was toward instructivism [χ^2 (1, $N = 142$) = 0.36, $p = 0.54$]. Also the results of Chi-square test showed there were no significant differences between the dominant aspect of Australian and American lecturers about this question [χ^2 (1, $N = 39$) = 0.37, $p = 0.54$] which was toward instructivism. The results of comparing the dominant aspect of administrative staff about learning goals indicated that the majority of administrative staff in both countries were oriented toward instructivism [χ^2 (1, $N = 24$) = 0.12, $p = 0.72$].

3.3.2 *Experiential Value*

The factor of experiential value with two orientations, namely abstract and concrete, has been measured by three questions. The orientation of abstract indicates an instructivism perspective and the orientation of concrete indicates a constructivism aspect of experiential value. The first question is about the congruence of learning with a real environment, the second question is about the outcome orientations of learning and the last question is about the practicality of learning.

3.3.2.1 Question 4: Congruence of Learning with Reality

The results in Australia showed that 71.4% of students ($n = 50$), 55.6% of lecturers ($n = 10$) and 54.5% of administrative staff ($n = 6$) explained that students learn from any kind of example as long as it makes sense. However, 28.6% of students ($n = 20$), 44.4% of lecturers ($n = 8$) and 45.5% of administrative staff ($n = 5$) claimed that students learn from examples as long as they are related to their work or personal life. This means that regardless of academic positions of participants the abstract dominant aspect of participants of one faculty in an Australian university about question 4 is instructivism [χ^2 (2, $N = 99$) = 2.44, $p = 0.29$].

The results in America about question 4 show that 66.7% of students ($n = 48$), 71.4% of lecturers ($n = 15$) and 46.2% of administrative staff ($n = 6$) believed in an abstract congruence of learning with reality. However, 33.3% of students ($n = 24$), 28.6% of lecturers ($n = 6$) and 53.8% of administrative staff ($n = 7$) claimed that there is concrete consistency between what people learn and how they apply that learning in their real life. As has been revealed, two-thirds of American students and lecturers believed in the abstract contingency of learning with reality which shows the dominant aspect of instructivism. However, it seems that more than half of administrative staff believed in the concrete consistency of learning with reality

Table 3.6 Comparison of responses to congruence of learning with reality

Paradigm	Students		Lecturers		Staff		χ^2
	N	%	*N*	%	*N*	%	
AUS participants							
Abstract	50	71.4	10	55.6	6	54.5	2.44
Concrete	20	28.6	8	44.4	5	45.5	
USA participants							
Abstract	50	66.7	15	71.4	6	46.2	2.93
Concrete	24	33.3	6	28.6	5	45.5	

Abstract = Instructivism; Concrete = Constructivism

which shows that the dominant aspect for them is toward constructivism. These differences in the results to question four between students and lecturers on the one hand and administrative staff on the other hand reflect the fact that students and lecturers more than administrative staff have been engaged with the consistency of learning with reality. Indeed they may believe that students learn from any kind of example as long as it makes sense even when it is not related to their work or personal life. However, based on the results of Chi-square test, there were no significant differences between students, lecturers and administrative staff's perspectives about question 4 [χ^2 (2, $N = 106$) = 2.93, $p = 0.23$] (see Table 3.6).

The outcome of comparing the results of Australians and Americans to question 4 revealed that the dominant aspect in both countries in students [χ^2 (1, $N = 142$) = 0.06, $p = 0.79$] and in lecturers [χ^2 (1, $N = 39$) = 1.06, $p = 0.30$] was toward instructivism. However, though the dominant aspect about question 4 in Australian administrators is also objectivism, in American administrators there is a constructivism perspective as the dominant aspect. Based on the Chi-square test, there were no significant differences between Australian and American students' perspectives on this question. Also according to the results of Chi-square test, there were no significant differences in the dominant aspect between Australian and American administrative staff on this question [χ^2 (1, $N = 24$) = 0.16, $p = 0.68$].

3.3.2.2 Question 5: Outcome Orientations of Learning

The results of question 5 about outcomes of learning in Australia revealed that 67.1% of students ($n = 47$), 61.1% of lecturers ($n = 11$) and 72.7% of administrative staff ($n = 8$) indicated that students have learned something when they can perform the activities requested by the instructor or course designer. However, 32.9% of students ($n = 23$), 38.9% of lecturers ($n = 7$) and 27.3% of administrative staff ($n = 3$) explained that students have learned some things when they have applied what they have learned to their everyday activities. Indeed the dominant aspect of all participants' answers regardless of their academic positions to this question is abstract, which indicates the traditional approach to the outcome of their learning [χ^2 (2, $N = 99$) = 1.12, $p = 0.56$].

Also, in Table 3.7, the results of this question to Americans revealed that 44.4% of students ($n = 32$), 47.6% of lecturers ($n = 10$) and 30.8% of administrative staff

Table 3.7 Comparison of responses to outcome orientations of learning

Paradigm	Students		Lecturers		Staff		χ^2
	N	%	*N*	%	*N*	%	
AUS participants							
Abstract	47	67.1	11	61.1	8	72.7	1.12
Concrete	23	32.9	7	38.9	3	27.3	
USA participants							
Abstract	32	44.4	10	47.6	4	30.8	0.96
Concrete	40	55.6	11	52.4	9	69.2	

Abstract = Instructivism; Concrete = Constructivism

(n = 4) indicated abstract outcomes of learning. However, 55.6% of American students (n = 40), 52.4% of lecturers (n = 11) and 69.2% of administrative staff (n = 9) indicated concrete outcomes of learning. Indeed the dominant aspect of participants' answers about the outcomes orientation of learning in America is constructivism [χ^2 (2, N = 106) = 0.96, p = 0.61].

Comparing the Australian and American results in this question shows that there were significant differences between the dominant aspect of Australian (instructivism) and American (constructivism) students to this question [χ^2 (1, N = 142) = 8.31, p = 0.004]. Also the results of Chi-square test show that there were significant differences between the dominant aspect of Australian (instructivism) and American (constructivism) administrative staff on this question [χ^2 (1, N = 24) = 4.19, p = 0.04]. However, there were no significant differences between the dominant aspect of Australian and American lecturers about outcomes of learning [χ^2 (1, N = 39) = 0.24, p = 0.62].

3.3.2.3 Question 6: Practicality of Learning

Concerning the sixth question which is about practicality of learning, the results for participants of one faculty in an Australian university showed that 76.6% of students (n = 55), 61.1% of lecturers (n = 11) and 63.6% of administrative staff (n = 7) believed in an abstract perspective in which they claimed that students are not expected to relate learning resources to their past or potential experiences. However, 21.4% of students (n = 15), 38.9% of lecturers (n = 7) and 36.4% of administrative staff (n = 4) believed in a concrete perspective to this question indicating that students are encouraged to apply 'knowledge' of learning to their activities at work and thus, are expected to learn from the actualization of those experiences. As a conclusion, the dominant aspect of all participants, regardless of their academic positions, in relation to practicality of learning is toward instructivism [χ^2 (2, N = 99) = 2.90, p = 0.23]. This may be due to the fact that even during engagement with an e-learning educational system, there is a traditional approach to the experimental value of the knowledge that has been learned during participation.

Concerning this question in American participants, results showed that 27.8% of students (n = 20), 33.3% of lecturers (n = 7) and 38.5% of administrative staff (n = 5) believed in an abstract perspective. However, 72.2% of students (n = 52),

Table 3.8 Comparison of responses to practicality of learning

Paradigm	Students		Lecturers		Staff		χ^2
	N	%	*N*	%	*N*	%	
AUS participants							
Abstract	55	76.6	11	61.1	7	63.6	2.90
Concrete	15	21.4	7	38.9	4	36.4	
USA participants							
Abstract	20	27.8	7	33.3	5	38.5	0.98
Concrete	52	72.2	14	66.7	8	61.5	

Abstract = Instructivism; Concrete = Constructivism

66.7% of lecturers (n = 14) and 61.5% of administrative staff (n = 8) indicated a concrete aspect to this question. As more than half of the participants of one faculty in an American in all academic positions claimed a concrete orientation of practicality of learning, the dominant orientation of participants to this question is constructivism [χ^2 (2, N = 106) = 0.98, p = 0.61] (see Table 3.8).

Comparing the results of question 6 between one faculty in an Australian university and one faculty in a US university reveals that there were significant differences between the dominant aspect of Australian (instructivism) and American (constructivism) students to this question [χ^2 (1, N = 142) = 38.72, p = 0.00]. Based on the Chi-square test results, there were no significant differences between the dominant aspect of Australian and American lecturers on this question [χ^2 (1, N = 39) = 3.00, p = 0.08]. In addition, the results indicated that there were no significant differences between Australian and American administrative staff on this question [χ^2 (1, N = 24) = 1.51, p = 0.21].

3.3.3 *Role of Instructor*

The factor of the role of instructor with two orientations, namely didactic and facilitative, has been measured by two questions. The orientation of didactic indicates an instructivism perspective and the orientation of facilitative indicates a constructivism perspective. The first question (Q7) is about the role of instructor related to learning needs and the second question (Q8) is about the role of instructor related to the source of learning.

3.3.3.1 Question 7: Role of Instructor Related to Learning Needs

According to Table 3.9, the results of question 7 for participants of one faculty in an Australian university showed that 81.4% of students (n = 57), 83.3% of lecturers (n = 15) and 72.7% of administrative staff (n = 8) believed in the didactic role of instructor related to learning needs in their e-learning system claiming that students follow a path of learning determined by the instructor because the instructor usually

Table 3.9 Comparison of responses to role of instructor related to learning needs

Paradigm	Students		Lecturers		Staff		χ^2
	N	%	*N*	%	*N*	%	
AUS participants							
Didactic	57	81.4	15	83.3	8	72.7	0.50
Facilitative	13	18.6	3	16.7	3	27.3	
USA participants							
Didactic	46	63.9	13	61.9	5	38.5	3.00
Facilitative	26	36.1	8	38.1	8	61.5	

Didactic = Instructivism; Facilitative = Constructivism

knows what students need to learn. However, 18.6% of students (n = 13), 16.7% of lecturers (n = 3) and 27.3% of administrative staff (n = 3) believed in a facilitative role of instructor related to learning needs, claiming that students learn from examples as long as they are related to their work or personal life. Based on these results the dominant aspect of Australians regardless of their academic positions about this question is toward instructivism [χ^2 (2, N = 99) = 0.50, p = 0.77].

Also, the results of question 7 for the participants of one faculty in an American university showed that 63.9% of students (n = 46), 61.9% of lecturers (n = 13) and 38.5% of administrative staff (n = 5) believed in the didactic role of instructor related to learning needs. However, 36.1% of students (n = 26), 38.1% of lecturers (n = 8) and 61.5% of administrative staff (n = 8) believed in the facilitative role of instructor related to learning needs. Although the dominant aspect of students and lecturers to this question is toward instructivism, it seems that the dominant aspect of administrative staff to this question is toward constructivism. The results of Chi-square test showed that there were no significant differences between the dominant aspect of students, lecturers and administrative staff in America to this question [χ^2 (2, N = 106) = 3.00, p = 0.22].

Comparing the results of Australians and Americans to this question revealed that there were significant differences between the dominant aspect of Australian (instructivism) and American (constructivism) students to this question [χ^2 (1, N = 142) = 4.01, p = 0.04]. Further the results of Chi-square test showed that there were no significant differences between the dominant aspect Australian and American lecturers on this question [χ^2 (1, N = 39) = 1.14, p = 0.28]. In addition, in the results of Chi-square test, there were no significant differences between the dominant aspects of Australian and American administrative staff to this question [χ^2 (1, N = 24) = 2.81, p = 0.09].

3.3.3.2 Question 8: Role of Instructor Related to the Source of Learning

Concerning question 8, the Australian response showed that 72.9% of students (n = 51), 66.7% of lecturers (n = 12) and 63.6% of administrative staff (n = 7) believed that students are taught by an "expert/source of knowledge" in the field about what they need to learn. However, 27.1% of students (n = 19), 33.6% of

Table 3.10 Comparison of responses to role of instructor related to the source of learning

Paradigm	Students		Lecturers		Staff		χ^2
	N	%	*N*	%	*N*	%	
AUS participants							
Didactic	51	72.9	12	66.7	7	63.6	0.56
Facilitative	19	27.1	6	33.6	4	36.4	
USA participants							
Didactic	50	69.4	9	42.9	5	38.5	7.77*
Facilitative	22	30.6	12	57.1	8	61.5	

Didactic = Instructivism; Facilitative = Constructivism
$*p < 0.05$

lecturers ($n = 6$) and 36.4% of administrative staff ($n = 4$) believed students are guided by an instructor who facilitates and shows them how to learn what they need to learn. As can be seen in Table 3.10, two-thirds of Australian participants' perspectives, regardless of their academic positions, about the role of instructor in their e-learning educational system are oriented towards didactic (instructivism) rather than facilitative [χ^2 (2, $N = 99$) = 0.56, $p = 0.75$]. This may be due to the fact that the traditional educational system has taught participants of one faculty in an Australian university that it is not you but the instructor who knows what you need to learn which came from the idea that one should quietly absorb knowledge handed over rather than obtain it oneself through the process of participation.

Concerning question 8, the American response showed that 69.4% of students ($n = 50$), 42.9% of lecturers ($n = 9$) and 38.5% of administrative staff ($n = 5$) believed in a didactic dimension of the role of instructor related to learning source. However, 30.6% of students ($n = 22$), 57.1% of lecturers ($n = 12$) and 61.5% of administrative staff ($n = 8$) believed in a facilitative dimension of the role of instructor related to learning source. Based on the Chi-square test, there were significant differences between the dominant aspect of students (instructivism), lecturers (constructivism) and administrative staff (constructivism) in America in the response to this question [χ^2 (2, $N = 106$) = 7.77, $p = 0.02$] (see Table 3.9).

Comparing the responses of Australians and Americans to question 8 revealed that the dominant orientation of students in both countries about the role of instructor related to learning source is toward instructivism [χ^2 (1, $N = 142$) = 0.20, $p = 0.65$]. Also the results of Chi-square test showed that there were no significant differences between the dominant aspects of Australian and American lecturers to this question [χ^2 (1, $N = 39$) = 2.21, $p = 0.13$]. Moreover, the results indicated that there were no significant differences between the dominant aspects of Australian and American administrative staff to this question [χ^2 (1, $N = 24$) = 1.51, $p = 0.21$].

3.3.4 *Value of Errors*

The factor of the value of errors with two dimensions, namely errorless learning and learning from experience, has been measured by two questions. The dimension of errorless learning indicates an instructivism orientation and the dimension of learning from experience indicates a constructivism orientation. The first question (Q9) is about errors in the process of learning and the second question (Q10) is about satisfaction of course designer with the learning.

3.3.4.1 Question 9: Errors in the Process of Learning

As can be seen in Table 3.11 regarding question 9, 40% of Australian students (n = 28), 38.9% of lecturers (n = 7) and 27.3% of administrative staff (n = 3) believed in errorless learning in which they claimed that *students learn until they make no errors on a test in the learning procedure*. However, 60% of Australian students (n = 42), 61.1% of lecturers (n = 11) and 72.7% of administrative staff (n = 8) believed in learning from experience which they explained as *students learn from their errors by experimenting with what they have learned.* Based on these results the dominant aspect of all Australians regardless of their academic position is toward constructivism in answer to this question [χ^2 (2, $N = 99$) = 0.81, $p = 0.66$].

Also, considering question 9, the American sample showed that 26.4% students (n = 19), 28.6% of lecturers (n = 6) and 7.7% of administrative staff (n = 1) believed in errorless learning in their e-learning system. However, 73.6% of American students (n = 53), 71.4% of lecturers (n = 15) and 92.3% of administrative staff (n = 12) believed in learning from experience which shows that the dominant aspect of all Americans regardless of their academic position is toward constructivism [χ^2 (2, $N = 106$) = 2.47, $p = 0.29$].

Comparing the results of question 9 between one faculty in an Australian university and one faculty in a US university responses showed that in both faculties the dominant orientation of all academic positions including students [χ^2 (1, $N = 142$) = 2.36, $p = 0.12$], lecturers [χ^2 (1, $N = 39$) = 0.10, $p = 0.74$] and administrative staff [χ^2 (1, $N = 24$) = 1.64, $p = 0.20$] is toward constructivism. The fact that the statement of question 9 includes an errorless learning orientation may have guided all participants to rate it poorly also should be considered. Today even in a traditional educational system,

Table 3.11 Comparison of responses to errors in the process of learning

Paradigm	Students		Lecturers		Staff		χ^2
	N	%	N	%	N	%	
AUS participants							
Errorless learning	28	40	7	38.9	3	27.3	0.81
Learning from experience	42	60	11	61.1	8	72.7	
USA participants							
Errorless learning	19	26.4	6	28.6	1	7.7	2.47
Unrestricted	53	73.6	15	71.4	12	92.3	

Errorless learning = Instructivism; Learning from experience = Constructivism learning until making NO errors on the test is not accepted. Indeed the statement of this question may not reveal the instructivism orientation of participants in this point of view.

3.3.4.2 Question 10: Satisfaction of Course Designer with the Learning

As regards to question 10, Australian responses showed that 38.6% of students (n = 27), 27.8% of lecturers (n = 5) and 18.2% of administrative staff (n = 2) believed in errorless learning which they described as *the instructor or course designer is satisfied if students take (fulfill a course) a test without making mistakes*. However, 61.4% of students (n = 43), 72.2% of lecturers (n = 13) and 81.8% of administrative staff (n = 9) believed in learning from experience which they described as *the instructor or course designer is satisfied if students learn from their mistakes*. Indeed the dominant aspect of all participants regardless of their academic position is toward constructivism [χ^2 (2, N = 99) = 2.17, p = 0.33].

The results of American responses to this question showed that 19.4% of students (n = 14), 19.0% of lecturers (n = 4) and 15.4% of administrative staff (n = 2) believed in an errorless learning perspective. However, 80.6% of students (n = 58), 81.0% of lecturers (n = 17) and 84.6% of administrative staff (n = 11) believed in learning from experience which showed that the dominant aspect of all Americans is toward constructivism [χ^2 (2, N = 106) = 0.11, p = 0.94] (see Table 3.12).

Comparing the results of question 10 between one faculty in an Australian university and one faculty in a US university revealed that there were significant differences between the dominant aspects of Australian (instructivism) and American (constructivism) students to this question [χ^2 (1, N = 142) = 6.31, p = 0.01]. In the results of Chi-square test, there were no significant differences between the dominant aspect of Australian (constructivism) and American (constructivism) lecturers to this question [χ^2 (1, N = 39) = 0.41, p = 0.51]. In addition, the results showed that there were no significant differences between the dominant aspects of Australian (constructivism) and American (constructivism) administrative staff to this question [χ^2 (1, N = 24) = 0.03, p = 0.85].

Table 3.12 Comparison of responses to satisfaction of course designer with the learning

Paradigm	Students		Lecturers		Staff		χ^2
	N	%	N	%	N	%	
AUS participants							
Errorless learning	27	38.6	5	27.8	2	18.2	2.17
Learning from experience	43	61.4	13	72.2	9	81.8	
USA participants							
Errorless learning	14	19.4	4	19.0	2	15.4	0.11
Learning from experience	58	80.6	17	81.0	11	84.6	

Errorless learning = Instructivism; Learning from experience = Constructivism

3.3.5 *Origin of Motivation*

The factor of origin of motivation with two dimensions, namely extrinsic and intrinsic motivation, has been measured by three questions. The dimension of extrinsic motivation indicates an instructivism orientation and the dimension of intrinsic motivation indicates an constructivism orientation. The first question (Q11) is about requirements of learning, the second question (Q12) is about reasons of learning and the third question (Q13) is about preference of learning.

3.3.5.1 Question 11: Requirements of Learning

Responses to question 11 showed that 61.4% of Australian students ($n = 43$), 55.6% of lecturers ($n = 10$) and 45.5% of administrative staff ($n = 5$) believed that there is an extrinsic origin of motivation for requirements of learning in their e-learning system which they described as *students take e-learning courses when they are required to* (*to pass the course or take a degree*).However, 38.6% of students ($n = 27$), 44.4% of lecturers ($n = 8$) and 54.5% of administrative staff ($n = 6$) believed that there is an intrinsic origin of motivation for requirements of learning namely that *students take e-learning courses when they want to* (*are genuinely interested in learning new knowledge or skills*). The results indicated that the dominant aspect of students and lecturers to this question is toward instructivism, however, the dominant aspect of administrative staff to this question is toward constructivism. The results of Chi-square indicated that there were no significant differences between the dominant aspects of students, lecturers and administrative staff in Australia to this question [χ^2 (2, $N = 99$) = 1.08, $p = 0.58$].

Reviewing the responses of Americans to question 11 showed that 48.6% of students ($n = 35$), 19.0% of lecturers ($n = 4$) and 23.1% of administrative staff ($n = 3$) believed that there is an extrinsic origin of motivation for requirements of learning in their e-learning system. However, 51.4% of students ($n = 37$), 81.0% of lecturers ($n = 17$) and 76.9% of administrative staff ($n = 10$) believed that there is an intrinsic origin of motivation for requirements of learning which showed the dominant aspect of constructivism in lecturers and staff American participants. However, the students' results are fairly close together which means that half of them show constructivism and half of them show instructivism [χ^2 (2, $N = 106$) = 7.63, $p = 0.02$] (see Table 3.13).

Comparing the results of Australian and American responses to question 11 indicates that there were no significant differences between the dominant aspects of Australian (instructivism) and American (constructivism) students to this question [χ^2 (1, $N = 142$) = 2.35, $p = 0.12$]. In the results of Chi-square test, there were significant differences between the dominant aspect of Australian (instructivism) and American (constructivism) lecturers to this question [χ^2 (1, $N = 39$) = 5.61, $p = 0.01$]. In addition, the results indicated that there were no significant differences between the dominant aspect of Australian (constructivism) and American (constructivism) administrative staff to this question [χ^2 (1, $N = 24$) = 1.34, $p = 0.24$].

Table 3.13 Comparison of responses to requirements of learning

Paradigm	Students		Lecturers		Staff		χ^2
	N	%	*N*	%	*N*	%	
AUS participants							
Extrinsic	43	61.4	10	55.6	5	45.5	1.08
Intrinsic	27	38.6	8	44.4	6	54.5	
USA participants							
Extrinsic	35	48.6	4	19.0	3	23.1	7.63*
Intrinsic	37	51.4	17	81.0	10	76.9	

Extrinsic = Instructivism; Intrinsic = Constructivism
$*p < 0.05$

3.3.5.2 Question 12: Reasons for Learning

Concerning question 12, the Australian sample showed that 57.1% of students ($n = 40$), 61.1% of lecturers ($n = 11$) and 72.7% of administrative staff ($n = 8$) believed there are extrinsic reasons for taking e-learning programs in their e-learning educational system which they explained as *students taking e-learning programs because they have no other option* (*conventional programs*). However, 42.9% of students ($n = 30$), 38.9% of lecturers ($n = 7$) and 27.3% of administrative staff ($n = 3$) believed that there are intrinsic motivation reasons for taking e-learning programs in their e-learning educational system which they explained as *students taking e-learning programs based on their interests which accommodate their specific needs*. Based on these results the dominant aspect of all participants of one faculty in an Australian university to this question is toward instructivism [χ^2 (2, $N = 99$) = 0.98, $p = 0.61$] (see Table 3.14).

Referring to question 12, the American sample showed that 33.3% of students ($n = 24$), 66.7% of lecturers ($n = 7$) and 23.1% of administrative staff ($n = 3$) believed there are extrinsic reasons for taking e-learning programs in their e-learning educational system which they explained as *students taking e-learning programs because they have no other option* (*conventional programs*). However, 66.7% of students ($n = 48$), 33.3% of lecturers ($n = 14$) and 76.9% of administrative staff ($n = 10$) believed that there are intrinsic motivation reasons for taking e-learning programs in their e-learning educational system which they explained as

Table 3.14 Comparison of responses to reasons for learning

Paradigm	Students		Lecturers		Staff		χ^2
	N	%	*N*	%	*N*	%	
AUS participants							
Extrinsic	40	57.1	11	61.1	8	72.7	0.98
Intrinsic	30	42.9	7	38.9	3	27.3	
USA participants							
Extrinsic	24	33.3	7	66.7	3	23.1	0.55
Intrinsic	48	66.7	14	33.3	10	76.9	

Extrinsic = Instructivism; Intrinsic = Constructivism

students taking e-learning programs based on their interests which accommodate their specific needs. Accordingly the dominant aspect of students and administrative staff to this question is toward constructivism, however, the dominant aspect of lecturers to this question is toward instructivism. However, Chi-square test results indicated that there were no significant differences between students, lecturers and administrators' perspectives in America on this question [χ^2 (2, N = 106) = 0.55, p = 0.75].

Comparing the responses to question 12 between Australian (constructivism) and American (instructivism) participants showed that the dominant orientations for taking e-learning programs in Australian and American students [χ^2 (1, N = 142) = 8.12, p = 0.004] are significantly different. Also the results of Chi-square test show there were no significant differences between Australian (instructivism) and American (instructivism) lecturers to this question [χ^2 (1, N = 39) = 3.00, p = 0.08]. In addition, the results of Chi-square test showed there were significant differences between the dominant aspects of Australian (constructivism) and American (instructivism) administrative staff to this question [χ^2 (1, N = 24) = 5.91, p = 0.01].

3.3.5.3 Question 13: Preference of Learning

Concerning question 13, Australian responses showed that 64.3% of students (n = 45), 72.2% of lecturers (n = 13) and 72.7% of administrative staff (n = 8) believed that there are extrinsic preferences for taking e-learning programs which indicates that *students mostly take courses in which they are told what they need to learn.* However, 35.7% of students (n = 25), 27.8% of lecturers (n = 5) and 27.3% of administrative staff (n = 3) believed that there are intrinsic preferences for taking e-learning programs which indicates that *students mostly take (are allowed to take) courses in which they choose what they need to learn.* Based on the Chi-square test there were no significant differences between the dominant aspects of students, lecturers and administrative staff (instructivism) in Australia to this question [χ^2 (2, N = 99) = 0.61, p = .71].

Also the responses to question 13 in America showed that 28.3% of students (n = 30), 57.1% of lecturers (n = 12) and 76.9% of administrative staff (n = 10) believed that there are extrinsic preferences for taking e-learning programs which indicates that *students mostly take courses in which they are told what they need to learn.* However, 35.7% of students (n = 25), 27.8% of lecturers (n = 5) and 27.3% of administrative staff (n = 3) believed that there are intrinsic preferences for taking e-learning programs which indicates that *students mostly take (are allowed to take) courses in which they choose what they need to learn.* Based on the Chi-square test, there were no significant differences between the dominant aspects of students, lecturers and administrative staff (instructivism) in America to this question [χ^2 (2, N = 106) = 6.16, p = 0.04] (see Table 3.15).

Comparing the response to question 13 between participants of one faculty in an Australian university and one faculty in a US university shows that there is similarity in orientation of instructivism in both countries in all academic positions

Table 3.15 Comparison of responses to preference of learning

Paradigm	Students		Lecturers		Staff		χ^2
	N	%	*N*	%	*N*	%	
AUS participants							
Extrinsic	45	64.3	13	72.2	8	72.7	0.61
Intrinsic	25	35.7	5	27.8	3	27.3	
USA participants							
Extrinsic	30	28.3	12	57.1	10	76.9	6.16*
Intrinsic	42	39.7	9	42.9	3	23.1	

Extrinsic = Instructivism; Intrinsic = Constructivism
*$p < 0.05$

regarding their extrinsic preference of learning [Students: χ^2 (1, $N = 142$) = 7.28, $p = 0.007$; lecturers: χ^2 (1, $N = 39$) = 0.95, $p = 0.32$ and administrative staff: χ^2 (1, $N = 24$) = 0.56, $p = 0.81$]. This similarity in instructivism orientation means that in both countries the preferences of students for taking courses are not considered by the e-learning system. From this perspective, there is a dominant traditional aspect even in the e-learning system about the *why* of learning, in which both students and instructors are following the way that has been defined by the system. This may lead to the conclusion that in such learning systems, students just receive graduation but do not acquire knowledge.

3.3.6 *Accommodation of Individual Differences*

The factor of accommodation of individual differences with two dimensions, namely non-existent and multifaceted, has been measured by two questions. The dimension of non-existent indicates the instructivism orientation and the dimension of multifaceted indicates the constructivism orientation. The first question (Q14) is about learning activities and the second question (Q15) is about consideration of needs and interests in learning.

3.3.6.1 Question 14: Learning Activities

As can be seen in Table 3.16, the responses to question 14 in the Australian sample showed that 74.3% of students ($n = 52$), 61.1% of lecturers ($n = 11$) and 90.9% of administrative staff ($n = 10$) believed that learning activities which can accommodate individual differences are non-existent. From this perspective, mostly they claimed that *the instructor or course designer uses very few learning activities and methods which allow students to learn through predetermined methods.* However, 25.7% of students ($n = 18$), 38.9% of lecturers ($n = 7$) and 9.1% of administrative staff ($n = 1$) believed that there are multifaceted learning activities which can accommodate individual differences. They claimed that *the instructor or course designer uses a variety of learning activities and instructional methods (like*

Table 3.16 Comparison of responses to learning activities

Paradigm	Students		Lecturers		Staff		χ^2
	N	%	*N*	%	*N*	%	
AUS participants							
Non-existent	52	74.3	11	61.1	10	90.9	3.16
Multifaceted	18	25.7	7	38.9	1	9.1	
USA participants							
Existent	28	38.9	12	57.1	5	38.5	2.31
Multifaceted	44	61.1	9	42.9	8	61.5	

Non-existent = Instructivism; Multifaceted = Constructivism

problem solving, case analysing, participation, etc.), *so that students can utilize what most suits their affect and their preferences.* Based on the Chi-square test there were no significant differences between the dominant aspect (instructivism) of students, lecturers and administrative staff in Australia on this question [χ^2 (2, $N = 99$) = 3.16, $p = 0.20$].

The responses to question 14 in the American sample showed that 38.9% of students ($n = 28$), 57.1% of lecturers ($n = 12$) and 38.5% of administrative staff ($n = 5$) believed that learning activities that can accommodate individual differences are non-existent. They claimed that *the instructor or course designer uses very few learning activities and methods which allow students to learn through predetermined methods.* However, 61.1% of students ($n = 44$), 42.9% of lecturers ($n = 9$) and 61.5% of administrative staff ($n = 8$) believed that there are multifaceted learning activities which can accommodate individual differences. They claimed that *the instructor or course designer uses a variety of learning activities and instructional methods (like problem solving, case analysing, participation,* etc.), *so that students can utilize what most suits their affect and their preferences.* Based on these results, although the dominant aspect of students and administrative staff is toward constructivism, the dominant aspect of lecturers is toward instructivism [χ^2 (2, $N = 106$) = 2.31, $p = 0.31$] (see Table 3.16).

Comparing the responses to question 14 between one faculty in an Australian university and one faculty in a USA university illustrated that the dominant orientation of Australian (instructivism) and American (constructivism) students to this question is significantly different [χ^2 (1, $N = 142$) = 18.07, $p = 0.00$]. However, the dominant orientation of both Australian and American lecturers is toward instructivism [χ^2 (1, $N = 39$) = 0.06, $p = 0.80$]. The results of Chi-square test showed that there were significant differences between the dominant aspect of Australian (instructivism) and American (constructivism) administrative staff on this question [χ^2 (1, $N = 24$) = 6.99, $p = 0.008$].

3.3.6.2 Question 15: Consideration of Needs and Interests in Learning

Concerning question 15, the Australian sample showed that 68.6% of students ($n = 48$), 61.1% of lecturers ($n = 11$) and 81.8% of administrative staff ($n = 9$)

believed that consideration of needs and interests which can accommodate individual differences is non-existent. Accordingly, they explained that *students' interests and needs are usually not considered in designing and providing courses (learning resources)*. However, 31.4% of students ($n = 22$), 38.9% of lecturers ($n = 7$) and 18.2% of administrative staff ($n = 2$) believed there is multifaceted consideration of needs and interests which can accommodate individual differences. They claimed that *students' needs and preferences are usually considered in designing and providing courses (learning resources)*. Based on the Chi-square test there were no significant differences between the dominant aspect of students, lecturers and administrative staff (instructivism) in Australia to this question [χ^2 (2, $N = 99$) = 1.36, $p = 0.50$].

Also the responses to question 15 in the American sample showed that 56.9% of students ($n = 41$), 52.4% of lecturers ($n = 11$) and 38.5% of administrative staff ($n = 5$) believed that consideration of needs and interests which can accommodate individual differences is non-existent. They explained that *students' interests and needs are usually not considered in designing and providing courses (learning resources)*. However, 43.1% of students ($n = 31$), 47.6% of lecturers ($n = 10$) and 61.5% of administrative staff ($n = 8$) believed there is multifaceted consideration of needs and interests which can accommodate individual differences. They claimed that *students' needs and preferences are usually considered in designing and providing courses (learning resources)*. Based on the Chi-square test, there were no significant differences between students, lecturers and administrative staff's perspectives in America on this question [χ^2 (2, $N = 106$) = 1.53, $p = 0.46$] (see Table 3.17).

Comparing the results of question 15 between participants in one faculty in an Australian university and one faculty in a US university revealed that there were no significant differences between Australian and American students' perspectives [χ^2 (1, $N = 142$) = 1.20, $p = 0.27$] and lecturers' perspectives to this question [χ^2 (1, $N = 39$) = 0.30, $p = 0.58$] which were toward instructivism. However, the results of Chi-square test showed there were significant differences between the dominant aspect of Australian (instructivism) and American (constructivism) administrative staff [χ^2 (1, $N = 24$) = 4.60, $p = 0.03$] to this question.

Table 3.17 Comparison of responses to consideration of needs and interests in learning

Paradigm	Students		Lecturers		Staff		χ^2
	N	%	N	%	N	%	
AUS participants							
Non-existent	48	68.6	11	61.1	9	81.8	1.36
Multifaceted	22	31.4	7	38.9	2	18.2	
USA participants							
Existent	41	56.9	11	52.4	5	38.5	1.53
Multifaceted	31	43.1	10	47.6	8	61.5	

Non-existent = Instructivism; Multifaceted = Constructivism

3.3.7 *Learner Control*

The factor of learner control with two dimensions, namely non-existent and unrestricted has been measured by two questions. The dimension of non-existence of learner control indicates the orientation of instructivism and the dimension of unrestricted learner control indicates the orientation of constructivism. The first question (Q16) is about limitations in learning and the second question (Q17) is about source of learning materials. Tables 3.18 and 3.19 show the results.

3.3.7.1 Question 16: Limitations in Learning

As can be seen in Table 3.18, the results of Q16 in Australia showed that 85.7% of students ($n = 60$), 72.2% of lecturers ($n = 13$) and 90.9% of administrative staff ($n = 10$) believed that learner control is non-existent claiming that *students are usually given a deadline or timed activities.* Indeed more than two-thirds of participants explained that the deadline for each learning activity is set by the system and is not under control of the students. However, 14.3% of students ($n = 10$), 27.8% of lecturers ($n = 5$) and 9.1% of administrative staff ($n = 1$) believed that *students can control the pace of their learning.* Based on the Chi-square test there were no significant differences between students, lecturers and administrative staff's perspectives (instructivism) in Australia on this question [χ^2 (2, $N = 99$) = 2.38, $p = 0.30$].

Table 3.18 Comparison of responses to limitations in learning

Paradigm	Students		Lecturers		Staff		χ^2
	N	%	*N*	%	*N*	%	
AUS participants							
Non-existent	60	85.7	13	72.2	10	90.9	2.38
Unrestricted	10	14.3	5	27.8	1	9.1	
USA participants							
Non-existent	47	65.3	16	76.2	9	69.2	1.14
Unrestricted	25	34.7	5	23.8	4	30.8	

Non-existent = Instructivism; Unrestricted = Constructivism

Table 3.19 Comparison of responses to source of learning materials

Paradigm	Students		Lecturers		Staff		χ^2
	N	%	*N*	%	*N*	%	
AUS participants							
Non-existent	60	85.7	15	83.3	9	81.8	0.15
Unrestricted	10	14.3	3	16.7	2	18.2	
USA participants							
Non-existent	54	75.0	14	66.7	9	69.2	0.65
Unrestricted	18	25.0	7	33.3	4	30.8	

Non-existent = Instructivism; Unrestricted = Constructivism

The answers to Q16 in the American sample showed that 65.3% of students ($n = 47$), 76.2% of lecturers ($n = 16$) and 69.2% of administrative staff ($n = 9$) believed that learner control is non-existent in their university claiming that *students are usually given a deadline or timed activities.* However, 34.7% of students ($n = 25$), 23.8% of lecturers ($n = 5$) and 30.8% of administrative staff ($n = 4$) believed that *students can control the pace of their learning.* Based on the Chi-square test, there were no significant differences between students, lecturers and administrative staff's perspectives in America (instructivism) on this question [χ^2 (2, $N = 106$) = 1.14, $p = 0.56$].

Comparing the results of question 16 between one faculty in an Australian university and one faculty in a US university revealed that there were significant differences between Australian and American students' perspectives on this question [χ^2 (1, $N = 142$) = 7.98, $p = 0.005$]. Also the results of Chi-square test showed there were no significant differences between the dominant aspect of Australian (instructivism) and American (instructivism) lecturers on this question [χ^2 (1, $N = 39$) = 0.08, $p = 0.77$]. In addition, the results showed there were no significant differences between Australian (instructivism) and American (instructivism) administrative staff to this question [χ^2 (1, $N = 24$) = 1.69, $p = 0.19$]. There is similarity regarding the dominant orientation of instructivism about limitations in learning activities, showing that in both countries, most participants believed that the deadline for learning activities is set by the system, not the students.

3.3.7.2 Question 17: Source of Learning Materials

Concerning question 17, the Australian sample showed that 85.7% of students ($n = 60$), 83.3% of lecturers ($n = 15$) and 81.8% of administrative staff ($n = 9$) believed that *the course features (the types of technologies included in the course, like chat, simulations) that will help students learn the materials are chosen by the instructor or course designer* but not by the students. However, 14.3% of students ($n = 10$), 16.7% of lecturers ($n = 3$) and 18.2% of administrative staff ($n = 2$) believed that *the course features that will help students learn the intended materials are chosen by students or with their contribution.* Based on the Chi-square test there were no significant differences between students, lecturers and administrative staff's perspectives in Australia on this question [χ^2 (2, $N = 99$) = 0.15, $p = .92$] toward instructivism.

Concerning question 17, the American sample showed that 75.0% of students ($n = 54$), 66.7% of lecturers ($n = 14$) and 69.2% of administrative staff ($n = 9$) believed that *the course features (the types of technologies included in the course, like chat, simulations) that will help students learn the materials are chosen by the instructor or course designer* but not by the students. However, 25.0% of students ($n = 18$), 33.3% of lecturers ($n = 7$) and 30.8% of administrative staff ($n = 4$) believed that *the course features that will help students learn the intended materials are chosen by students or with their contribution.* Based on the Chi-square test, there were no significant differences between students, lecturers and administrative staff's perspectives in America on this question [χ^2 (2, N = 106) = 0.65, p = 0.72] toward instructivism (see Table 3.19).

Comparing the responses to question 17 between one faculty in an Australian university and one faculty in a US university about the source of learning materials showed that there is still similarity in orientation of instructivism in both countries. From this perspective, the majority of students claimed the instructor centre rather than student centre was the source of choosing materials [χ^2 (1, N = 142) = 2.57, p = .10]. Also the results of Chi-square test showed that there were no significant differences between Australian and American lecturers' perspectives on this question [χ^2 (1, N = 39) = 1.41, p = 0.23] toward instructivism. In addition, the results of Chi-square test showed that there were no significant differences between Australian and American administrative staff's perspectives on this question [χ^2 (1, N = 24) = 0.50, p = 0.47] toward instructivism. Interestingly, the fact that the dominant approach of learner control is objectivism shows that there is strong restriction for learners in both e-learning systems.

3.3.8 *User Activity*

The factor of user activity with two dimensions, mathemagenic and generative has been measured by two questions. The dimension of mathemagenic of user activity indicates the orientation of instructivism and the dimension of generative user activity indicates the orientation of constructivism. The first question (Q18) is about knowledge engagements and the second question (Q19) is about learning resources. Tables 3.20 and 3.21 show the results.

3.3.8.1 Question 18: Knowledge Engagements

According to Table 3.20, the results of the Australian sample regarding Q18 showed that 74.3% of students (n = 52), 77.8% of lecturers (n = 14) and 81.8% of administrative staff (n = 9) believed that *students do not have any involvement in producing and representing knowledge* which is the mathemagenic, passive

Table 3.20 Comparison of responses to knowledge engagements

Paradigm	Students		Lecturers		Staff		χ^2
	N	%	N	%	N	%	
AUS participants							
Mathemagenic	52	74.3	14	77.8	9	81.8	0.11
Generative	18	25.7	4	22.2	2	18.2	
USA participants							
Mathemagenic	36	50.0	10	47.6	6	46.2	0.08
Generative	36	50.0	11	52.4	7	53.8	

Mathemagenic = Instructivism; Generative = Constructivism

Table 3.21 Comparison of responses to learning resources

Paradigm	Students		Lecturers		Staff		χ^2
	N	%	*N*	%	*N*	%	
AUS participants							
Mathemagenic	46	65.7	9	50	7	63.6	1.51
Generative	24	34.3	9	50	4	36.4	
USA participants							
Mathemagenic	45	62.5	9	42.9	3	23.1	10.85*
Generative	27	37.5	12	57.1	10	76.9	

Mathemagenic = Instructivism; Generative = Constructivism
$^*p < 0.05$

approach to acquiring knowledge. However, 25.7% of students (n = 18), 22.2% of lecturers (n = 4) and 18.2% of administrative staff (n = 2) believed that *students are engaged in the process of creating, elaborating, or representing knowledge* which is the generative active approach in the process of learning. These results indicated that two-thirds of participants were oriented towards instructivism rather than constructivism. Based on the Chi-square test there were no significant differences between students, lecturers and administrative staff's perspectives in Australia on this question [χ^2 (2, N = 99) = 0.11, p = 0.94].

Also, the response to Q18 by Americans showed that 50.0% of students (n = 36), 47.6% of lecturers (n = 10) and 46.2% of administrative staff (n = 6) believed that *students do not have any involvement in producing and representing knowledge* which is the mathemagenic, passive approach to acquiring knowledge. However, 50.0% of students (n = 36), 52.4% of lecturers (n = 11) and 53.8% of administrative staff (n = 7) believed that *students are engaged in the process of creating, elaborating, representing of knowledge* which is the generative active approach in the process of learning. Based on the Chi-square test, there were no significant differences between students, lecturers and administrative staff's perspectives in America on this question [χ^2 (2, N = 106) = 0.08, p = 0.95].

Comparing results of question 18 between one faculty in an Australian university and one faculty in a US university showed that the dominant orientation of knowledge engagement in one faculty in an Australian university is instructivism which means that two thirds of students, lecturers and administrative staff believed that the students do not play a role in creating or producing knowledge. However, the responses of American students to this question revealed that half of them believed in an instructivism orientation and half of them believed in a constructivism orientation [χ^2 (1, N = 142) = 8.88, p = 0.003]. It can be concluded that some students have engaged in the process of learning by creating and producing their own learning materials however, some of them act passively in the process of learning. In addition more than half of American lecturers and administrative staff believed in a constructivism orientation of user activity in their e-learning system. It is worth noting that just one participant can change the pattern of dominant orientation in lecturers and administrative staff. Indeed replication of results may cause different results. Also according to the results of Chi-square test, there were no significant differences between Australian and American lecturers' perspectives on this question [χ^2 (1, $N = 39$) = 3.72, $p = 0.05$]. In addition, the results of Chi-square test showed there were no significant differences between Australian and American administrative staff's perspectives on this question [χ^2 (1, $N = 24$) = 1.73, $p = .18$].

3.3.8.2 Question 19: Learning Resources

Concerning question 19, Australian responses showed that 65.7% of students ($n = 46$), 50% of lecturers ($n = 9$) and 63.6% of administrative staff ($n = 7$) believed that *students usually access representations of provided learning resources according to a predetermined path.* However, 34.3% of students ($n = 24$), 50% of lecturers ($n = 9$) and 36.4% of administrative staff ($n = 4$) believed that *the learning resources of the course are usually presented to students, but they create their own uses of the learning resources within the course.* Although the perspective of lecturers to this question is half instructivism and half constructivism, students and administrative staff are strongly oriented towards objectivism. Based on the Chi-square test there were no significant differences between students, lecturers and administrative staff's perspectives in Australia on this question [χ^2 (2, $N = 99$) = 1.51, $p = 0.46$].

Concerning question 19, American responses showed that 62.5% of students ($n = 45$), 42.9% of lecturers ($n = 9$) and 23.1% of administrative staff ($n = 3$) believed in a mathemagenic source of learning claiming that *students usually access representations of provided learning resources according to a predetermined path.* However, 37.5% of students ($n = 27$), 57.1% of lecturers ($n = 12$) and 76.9% of administrative staff ($n = 10$) believed in a generative source of learning claiming that *the learning resources of the course are usually presented to students, but they create their own uses of the learning resources within the course.* Based on the Chi-square test, there were significant differences between students and lecturers with an instructivism orientation and administrative staff with a constructivism orientation in America on this question [χ^2 (2, $N = 106$) = 10.85, $p = 0.004$] (see Table 3.21).

Comparing the results of question 19 between participants of one faculty in an Australian university and participants of one faculty in an American university showed there were no significant differences between Australian and American students' perspectives on this question [χ^2 (1, $N = 142$) = 0.15, $p = 0.69$] toward instructivism. Also according to the results of Chi-square test, there were no significant differences between Australian and American lecturers' perspectives on this question [χ^2 (1, $N = 39$) = 0.19, $p = 0.65$] toward instructivism. In addition, the results of Chi-square test showed there were significant differences between Australian (instructivism) and American (constructivism) administrative staff on this question [χ^2 (1, $N = 24$) = 5.91, $p = 0.01$].

3.3.9 *Collaborative Learning*

The factor of collaborative learning with two dimensions, namely unsupported and integrated, was measured by two questions. The dimension of unsupported reflected the orientation of instructivism, however, the dimension of integrated reflected the orientation of constructivism. The first question (Q20) is about approach to learning activities and the second question (Q21) is about facilities of learning.

3.3.9.1 Question 20: Approach of Learning Activities

According to Table 3.22, the responses to Q20 by Australians showed that 58.6% of students (n = 41), 44.4% of lecturers (n = 8) and 63.6% of administrative staff (n = 7) believed that the approach to learning activities in their e-learning educational system is unsupported claiming *students usually work individually on their learning activities or projects.* However, 41.4% of students (n = 29), 55.6% of lecturers (n = 10) and 36.4% of administrative staff (n = 4) believed that the approach to learning activities in their e-learning educational system is integrated with others claiming *students usually (are encouraged to) work with a group on their learning activities or projects.* Based on the Chi-square test there were no significant differences between students, lecturers and administrative staff's perspectives in Australia on this question [χ^2 (2, N = 99) = 2.61, p = 0.27] toward instructivism.

Also, the responses to Q20 by Americans showed that 43.1% of students (n = 31), 14.3% of lecturers (n = 3) and 30.8% of administrative staff (n = 4) believed that the approach to learning activities in their e-learning educational system is unsupported claiming *students usually work individually on their learning activities or projects.* However, 56.9% of students (n = 41), 85.7% of lecturers (n = 18) and 69.2% of administrative staff (n = 9) believed that the approach to learning activities in their e-learning educational system is integrated. Based on the Chi-square test, there were significant differences between students, lecturers and administrative staff's perspectives in America on this question [χ^2 (2, N = 106) = 6.01, p = 0.04].

Table 3.22 Comparison of responses to approach of learning activities

Paradigm	Students		Lecturers		Staff		χ^2
	N	%	*N*	%	*N*	%	
AUS participants							
Unsupported	41	58.6	8	44.4	7	63.6	2.61
Integrated	29	41.4	10	55.6	4	36.4	
USA participants							
Unsupported	31	43.1	3	14.3	4	30.8	6.01*
Integrated	41	56.9	18	85.7	9	69.2	

Unsupported = Instructivism; Integrated = Constructivism
*$p < 0.05$

Comparing the responses to question 20 about the approach of learning activities between one faculty in an Australian university and one faculty in a US university showed that there were no significant differences between Australian and American students' perspectives on this question [χ^2 (1, $N = 142$) = 3.41, $p = 0.06$] toward instructivism. Also the results of Chi-square test showed there were significant differences between Australian and American lecturers' perspectives on this question [χ^2 (1, $N = 39$) = 4.35, $p = 0.03$]. In addition, according to the results of Chi-square test, there were no significant differences between Australian and American administrative staff's perspectives on this question [χ^2 (1, $N = 24$) = 0.08, $p = 0.77$].

3.3.9.2 Question 21: Facilities of Learning to Act Collaboratively

Concerning question 21, answers by Australians showed that 25.7% of students ($n = 18$), 33.3% of lecturers ($n = 6$) and 36.4% of administrative staff ($n = 4$) believed that *there are limited or no facilities (technical) for setting up collaborating learning in their learning environments.* However, 74.3% of students ($n = 52$), 66.7% of lecturers ($n = 12$) and 63.6% of administrative staff ($n = 7$) believed that *a wide range of different facilities and features are provided for setting up collaborating learning in their learning environments.* Based on the Chi-square test there were no significant differences between students, lecturers and administrative staff's perspectives in Australia on this question [χ^2 (2, $N = 99$) = 0.80, $p = 0.66$] toward constructivism.

Also, the answers to question 21 by Americans showed that 18.1% of students ($n = 13$), 19.0% of lecturers ($n = 4$) and 100% of administrative staff ($n = 13$) believed in unsupported facilities for learning claiming that *there are limited or no facilities (technical) for setting up collaborating learning in their learning environments.* However, 81.9% of students ($n = 59$), 81.0% of lecturers ($n = 17$) and no administrative staff ($n = 0$) believed in integrated facilities for learning claiming that *a wide range of different facilities and features are provided for setting up collaborating learning in their learning environments.* Based on the Chi-square test, there were no significant differences between students, lecturers and administrative staff's perspectives in America on this question [χ^2 (2, $N = 106$) = 2.84, $p = 0.24$] (see Table 3.23).

Table 3.23 Comparison of responses to facilities of learning to act collaboratively

Paradigm	Students		Lecturers		Staff		χ^2
	N	%	*N*	%	*N*	%	
AUS participants							
Unsupported	18	25.7	6	33.3	4	36.4	0.80
Integrated	52	74.3	12	66.7	7	63.6	
USA participants							
Unsupported	13	18.1	4	19.0	13	100	2.84
Integrated	59	81.9	17	81.0	0	0	

Unsupported = Instructivism; Generative = Constructivism

These results showed that the dominant orientation of participants of one faculty in an Australian university was constructivism. This indicates that the e-learning environment tries to support users with respect to different facilities. Similar results in one faculty in an American university indicated that in both countries the facilities for collaborative learning are constructed as good. Based on the Chi-square test, there were no significant differences between Australian and American students' perspectives on this question [χ^2 (1, $N = 142$) = 1.22, $p = 0.26$]. Also the results of Chi-square test showed there were no significant differences between Australian and American lecturers' perspectives on this question [χ^2 (1, $N = 39$) = 1.03, $p = 0.30$]. In addition, according to the results of Chi-square test, there were significant differences between Australian and American administrative staff's perspectives on this question [χ^2 (1, $N = 24$) = 5.67, $p = 0.01$].

As shown in Fig. 3.1, to sum up, the dominant aspect of the cultural dimension in one faculty in an Australian university in all academic positions is towards instructivism in most factors. The reason for that dominant orientation is that most participants of one faculty in an Australian university belonged to an international community (mostly students are from eastern Asian culture) rather than the domestic environment. This is congruent with the evidence that shows that the rate of international students who are attending e-learning courses in Australia is higher than domestic students.

The dominant aspect of constructivism in e-learning in America was focused on a student centred orientation which shows the level of development of the learning environment in America in both quality and quantity. This approach to e-learning reflected the stage of development as well as the cultural social background of the participants especially the students. The fact that America has been known as the

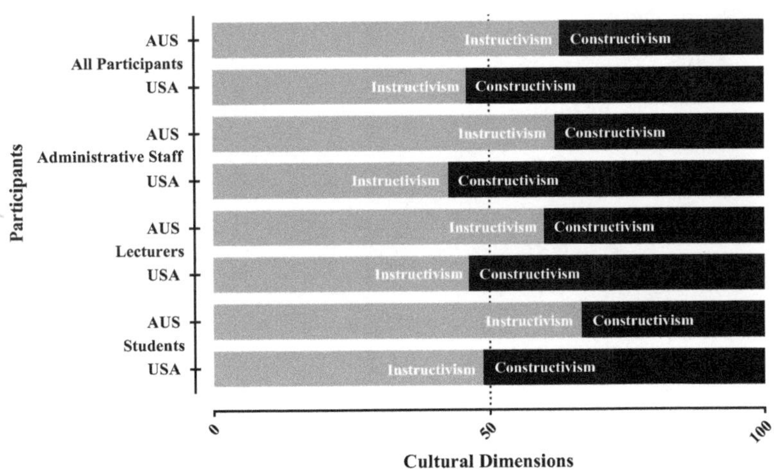

Fig. 3.1 Mean level of cultural dimensions based on participants in one faculty in an Australian and one faculty in a US university

most developed country in the world in relation to high technology can affect the e-learning educational perspective of its suppliers. For sure, the more highly novel the technological educational system, the stronger the perspective of constructivism in the academics.

This chapter discussed the dominant cultural orientations of e-practice in Australian and American e-learning systems and then compared the results of the two countries. The following chapter investigates the current status of e-practice such as pedagogy, performance appraisal, instructional design, and technological, administrative and support services in one faculty in an Australian and one faculty in a US university.

Chapter 4
The Current Status of E-Practice

4.1 Introduction

The principles of constructivist educational theory have come to be central to e-learning practice (Nkonge and Gueldenzoph 2006). Constructivist thought on e-pedagogy has provided basic principles of constructivism learning theory (Doolittle 1999; Hein 1991) for e-teaching best practices (Alley and Jansak 2001; Hacker and Niederhauser 2000; Keeton 2004). E-practice focused on learning and teaching processes is based on operational policies and practice standards for virtual learning environments (Kala et al. 2010). According to the evidence, practice which is based on learning and teaching theories can support online learning courses by developing a model for the learning and teaching process (Oliver 2001; Thurmond 2002).

A large number of researchers have directed their attention to the field of e-learning practice. These studies provide a variety of models, guidelines, critical success factors and benchmarks put forward as best e-learning practice in order to enhance and assure quality in higher education institutes. To give a comprehensive picture of the practice of e-learning, current e-practice work is reviewed and divided into two methodological approaches, namely theoretically oriented and practically oriented e-practice work. However, such a framework cannot include all the contributions to and research studies about e-practice; there may be other e-practice researches not presented in this research, for example, some e-practice work is focused on technologically-driven aims without any attempt at pedagogy-driven design. In light of the need for quality improvement of learning practice, investigation of comparative e-practices such as pedagogy, performance appraisal, instructional design, and technological, administrative and support services is critically important (Chickering et al. 1987; Commissions 2001; Dragon et al. 2013; Finger et al. 2006; FitzPatrick 2012; Granato 2008; Holsapple and Lee-Post 2006; Jacobson et al. 2005; Khan and Kala et al. 2010; Marshall 2012; Phipps and Merisotis 2000; Sangrà 2002; Selim 2007; Zhao 2003; Zhou 2012;).

S.H. Sadeghi, *E-Learning Practice in Higher Education: A Mixed-Method Comparative Analysis*, Studies in Systems, Decision and Control 122,
https://doi.org/10.1007/978-3-319-65939-8_4

Consequently, the second aim of this current research is to compare e-practice between an Australian university and an American university in order to ultimately improve the quality of online learning courses in Australian and American universities. Accordingly, a researcher questionnaire addressing e-practice factors was devised. To provide a more comprehensive understanding of these factors and sub-factors, they are summarized in Table 4.1.

Table 4.1 e-Practice factors and sub-factors

Factors	Sub-factors	References
Pedagogical practice	Student-centred interactivity Socio-communication Learning environment Assessment Learning resources	Chickering et al. (1987), Sangrà (2002), Finger et al. (2006), Khan (2008) and Zhou (2012)
Technological practice	The technological infrastructure Functionality of platforms Accessibility Reasonably User interface design	Sangrà (2002), Finger et al. (2006), Jacobson (2005) Holsapple and Lee-Post (2006), Volery and Lord (2000), Selim (2007), Khan (2008), FitzPatrick (2012) and Guribye (2015)
Instructional design practice	Clarifying expectations Personalisation Learning scenarios Organizing resources Quality and accuracy	Phipps and Merisotis (2000), Finger et al. (2006), Putnam and Borko (2000) and Marshall (2012)
Organisational practice	Institutional affairs Administrative affairs Research and development Precedent and reputation	Novak (2002), Sangrà (2002), Khan (2008) and Dragon et al. (2013)
Support services practice	Administrative support services Technical support services Academic support services	Phipps and Merisotis (2000), Sangrà (2002), Finger et al. (2006), Volery and Lord (2000), Selim (2007), Khan (2008) and FitzPatrick (2012)
Performance appraisal practice	Cost-effectiveness Learning Effectiveness Satisfaction	Phipps and Merisotis (2000), Sangrà (2002), Sloan-C (2002), Khan (2005), Kala et al. (2010), Zhou (2012) and FitzPatrick (2012)

4.2 Method

4.2.1 *Participants and Design*

The group participants in this study are basically the same as in the previous study described in Chap. 3 but some people were unable to participate in this study so the researcher added some new volunteers into this study. A total of 231 participants from an Australian university and an American university were recruited to take part in this study through an online invitation email asking for volunteers. To check the normality of the distribution and homogeneity of variance in this sample the researcher applied several tests including Boxplot and Kolmogorov–Smirnov. The results indicated 16 cases as outliers (7 Australian cases and 9 American cases) so they were excluded from the main analysis. Of the remaining sample of 215, 129 participants were female, and 86 were male. They reported their age as 20–30 years ($n = 99$), 30–40 years ($n = 68$) and 40–50 years ($n = 48$). Of them, 149 participants were students, 45 were lecturers and 21 were administrative staff. The type of previous experience of participants in e-learning educational system was categorised as both blended and online ($n = 155$) or fully online ($n = 60$). Table 4.2 is a summary of the demographic makeup of the participants of both countries.

The primary independent variables in this study were position of participants and the country. The dependent variables were pedagogical practice, technological practice, organizational practice, support practice, instructional design practice and performance appraisal practice.

4.2.2 *Material*

4.2.2.1 e-Practice Questionnaire

The instrument used was a questionnaire self-constructed by the researcher. Exploratory factor analysis was applied to test the validity of the constructed

Table 4.2 Demographic information based on country

Country	Gender	N	Age	N	Position	N	Experience	N
AUS	Female	59	20–30	57	Student	71	Blended and online	62
	Male	40	30–40	20	Lecturers	20	Fully online	37
	–	–	40–50	22	Staff	8	–	–
	Total	99						
USA	Female	70	20–30	42	Student	78	Blended and online	93
	Male	46	30–40	48	Lecturers	25	Fully online	23
	–	–	40–50	26	Staff	13	–	–
	Total	116						
All total	215							

questionnaire. The results of EFA showed that the Kaiser–Meyer–Olkin measure of sampling adequacy was equal to 0.84; this value is above the recommended value of 0.6. Also the results of Bartlett's Test of Sphericity were significant [χ^2 (903) = 3955.92, p = 0.000]. The communalities of items were above 0.4. Indeed, factor analysis was conducted with all 43 items.

Principle components analysis was applied to introduce and test the composite e-practice score. The result showed that the six e-practice factors loaded on only one factor of e-practice, with an eigenvalue of 21.378, which can explain 49.717% of the variance. To conclude, a one-factor solution was defined. In addition, internal consistency of the e-practice scales showed Cronbach's alpha as 0.88 which was in the high range of reliability. To empower the validity of the e-practice questionnaire further research would be suggested with a different population.

In the first part of the questionnaire, demographic information of participants was sought including their gender, age, position, the type of their e-learning experience (fully online or blended) as well as their country. The second part of the questionnaire had 43 questions to which participants provided their answers under the following instruction: "*As part of an international eLearning research project, the University of (blank) is collaborating with the University of (blank) to compare and contrast practices at the two institutions. Your responses will help to contribute to a broader understanding of similarities, differences, and best practices in the e-Learning area. We want to thank you in advance for agreeing to participate in this online survey.*"

Participants answered each question by using the Likert scale [1 = Extremely Poor, 2 = Poor, 3 = Average, 4 = Above Average (good), 5 = Excellent]. It is worth mentioning that three versions of the e-practice questionnaire were presented to participants based on their positions (see Appendix C). In each version the items were the same but the structure of each question was modified to be appropriate to the position of participants whether they were students, lecturers or administrative staff. To assess the *e-practice* experience of the participants, the researcher enquired into 6 main factors of e-practice, namely pedagogical e-practice, technological e-practice, instructional design e-practice, organisational e-practice, support e-practice and performance appraisal e-practice using 43 questions. They are presented below with the sub-factors, items, questions and range of scores. The factor of pedagogical e-practice has 5 sub factors elicited by 13 questions: student-centred interactivity, socio-communication, learning environment, assessment and learning resources (Chickering et al. 1987; Finger et al. 2006; Khan 2008; Sangrà 2002; Zhou 2012). The range of scores was 13–65 (Table 4.3).

The factor of *technological e-practice* has 5 sub factors elicited by 7 questions: technological development, functionality of platforms, accessibility, reasonable and user interface design (Finger et al. 2006; FitzPatrick 2012; Guribye 2015; Holsapple and Lee-Post 2006; Jacobson et al. 2005; Khan 2008; Sangrà 2002; Selim 2007; Volery and Lord 2000). The scores ranged from 7 to 35 (Table 4.4).

The factor of *instructional design e-practice* has 5 sub factors elicited by 8 questions: clarifying expectations, personalisation, learning scenarios, organizing resources, and accuracy and awareness (Finger et al. 2006; Marshall 2012; Phipps

Table 4.3 Sub-factors, items and questions of pedagogical e-practice

Sub-factors	Items	Questions
Student-centred interactivity	Student centred practices	1
	Interactive network classroom	5
	Using the blackboard discussion board	11
Socio-communication	Effective communication	16
	Facilities and opportunities	19
	The social interactive tools	20
	Competitive environment	24
Learning environment	Flexible environment system	6
	Environmental learning facilities	41
Assessment	Classroom constructive feedback	3
	Academic honesty plagiarism policy	10
	Feedback on assessment results	15
Learning resources	Access to e-resources	8

Table 4.4 Sub-factors, items and questions of technological e-practice

Sub-factors	Items	Questions
Technological development	Infrastructural development	25
	IT development	26
Functionality of platforms	LMS capabilities	31
	E-learning platforms	32
Accessibility	Accessibility practice	27
Reasonable	Reasonable	28
User interface design	User-friendly and versatile	33

Table 4.5 Sub-factors, items and questions of instructional design e-practice

Sub-factors	Items	Questions
Clarifying expectations	Clear objectives and expectations	7
	The outline and syllabus	35
Learning scenarios	The content of modules	2
	Effective instructional strategies	4
Quality and accuracy	Resources of instructional	22
	Reliable materials	23
Personalization	Personalization	21
Organizing rescores	Organizing online materials	9

and Merisotis 2000; Putnam and Borko 2000). The range of scores was 8–36 (Table 4.5).

The factor of *organisational e-practice* has 4 sub factors elicited by 4 questions: institutional affairs, administrative affairs, research and development (R&D) and

Table 4.6 Sub-factors, items and questions of organisational e-practice

Sub-factors	Items	Questions
Institutional affaires	Clear and effective practice	30
Administrative affairs	Supportive administrative practice	12
Research and development	Opportunities and facilities	18
Precedent and reputation	University e-learning reputation	13

Table 4.7 Sub-factors, items and questions of support services e-practice

Sub-factors	Items	Questions
Administrative support services	Control and accountability system	34
Technical support services	Helpdesk support	29
Academic support services	Academic administration support	14

Table 4.8 Sub-factors, items and questions of performance appraisal e-practice

Sub-factors	Item	Questions
Cost-effectiveness	Tuition rates	17
	Cost-effectiveness of online course	39
Learning effectiveness	Improving quality of outcome	36
	Effective feedback on performance	37
	Standards of performance practice	38
Satisfaction	Achieve satisfactory performance	40
	Motivation to achieve outcomes	42
	E-learning experience satisfaction	43

precedent and reputation (Dragon et al. 2013; Khan 2008; Novak 2002; Sangrà 2002). The scores ranged from 4 to 20 (Table 4.6).

The factor of *support services* practices includes 3 sub factors and is elicited by 3 questions: administrative support services, technical support services and academic support services (Finger et al. 2006; FitzPatrick 2012; Khan 2008; Sangrà 2002; Selim 2007; Phipps and Merisotis 2000; Volery and Lord 2000). The range of scores was 3–15 (Table 4.7).

The factor of *performance appraisal* has 3 sub factors namely cost-effectiveness, learning effectiveness and satisfaction (FitzPatrick 2012; Kala et al. 2010; Khan 2005; Phipps and Merisotis 2000; Sangrà 2002; Sloan-C 2002; Zhou 2012) elicited by 8 questions. The range of scores was 8–40 (Table 4.8).

4.2.3 *Procedure*

After providing ethical approval, the study was conducted by creating an online questionnaire of e-practice using Lime Survey software. The link to the questionnaire then was sent to the e-learning centre of health sciences in both universities. The e-learning coordinators of each university then sent the link of the survey to their lecturers, administrative staff and students who were engaged with online courses. The participants responded to the questionnaire voluntarily. They first provided demographic information, then they turned to the main questionnaire. The last page was a thank you page.

4.3 Results of Current Status of E-Practice

This section is concerned with the e-practice aspects of the two countries, Australia and the United States of America. The six main factors are dealt with in relation to the e-learning practices. Firstly the results of each factor and sub factor in relation to position in the Australian sample are discussed. Secondly the results of each individual factor in relation to this variable within the American sample are dealt with. Subsequently a comparison of the results of each individual factor within the American and Australian samples is presented.

4.3.1 Pedagogical E-Practice Factor Results

The pedagogical e-practice factor was measured by 5 sub factors namely: student centred interactivity, socio-communication, assessment, learning resources and learning environment. In this section, the results of each sub factor based on academic position and experience of participants in America and Australia about e-learning courses is reported. At the end the total results of all sub factors of the main factor of pedagogical e-practice have been reported.

4.3.1.1 Student Centred Interactivity

Table 4.9 reports the means and standard deviations of the student centred interactivity sub-factor based on academic positions of Australian participants. As can be seen in this table, the highest mean regarding this sub factor belonged to administrative staff ($M = 11.12$, $SD = 1.12$). After administrative staff, lecturers

Table 4.9 Mean, SD, and F value of evaluation of student centred interactivity

Country	Students		Lecturers		Staff		*F*	*P*
	M	*SD*	*M*	*SD*	*M*	*SD*		
AUS participants	9.07	1.45	9.70	2.27	11.12	1.12	6.22	0.003**
USA participants	10.39	1.13	11.10	1.01	11.23	1.01	6.21	0.003**

** $p < 0.01$

reported this factor next highest ($M = 9.70$, $SD = 2.27$) and the lowest score was reported by the students ($M = 9.07$, $SD = 1.45$). To investigate if there are any differences in the evaluation of student centred interactivity between students, lecturers and administrative staff, ANOVA was applied. The results showed that there was a significant effect of academic position on evaluation of student centred interactivity by participants of one faculty in an Australian university [F (2, 98) = 6.22, $p = 0.003$]. An LSD multiple comparison test between the three academic positions revealed that administrative staff reported this factor significantly higher than students and lecturers. However, the evaluation by lecturers and students of this sub-factor were the same. The results showed that all participants of one faculty in an Australian university believed that student centred interactivity e-practice is above average.

In one faculty in a US university, as can be seen in this table, the highest mean of answers to this sub factor belonged to administrative staff ($M = 11.23$, $SD = 1.01$). After administrative staff, the lecturers reported this factor ($M = 11.10$, $SD = 0.91$) with the next highest mean and the lowest score was reported by the students ($M = 10.39$, $SD = 1.13$). To investigate if there are any differences on evaluation of student centred interactivity between American students, lecturers and administrative staff, ANOVA was applied. The results showed that there was a significant main effect of academic position on evaluation of student centred interactivity by participants of one faculty in an American university [F (2, 115) = 6.21, $p = 0.003$]. An LSD multiple comparison test between the three levels of academic positions revealed that students reported this factor significantly lower than administrative staff and lecturers. However, the evaluation of this sub factor between lecturers and administrative staff was the same. The results showed that all participants of one faculty in a US university believed that student centred interactivity was above average.

Comparing the answers of participants of one faculty in an Australian university and one faculty in a US university showed that there were no significant differences on evaluation of this sub factor between Australian and American administrative staff; both evaluated this sub factor higher than students and lecturers [F (1, 20) = 0.05, $p = 0.82$]. Also the results of ANOVA revealed that there was significant difference in evaluation of this sub factor between Australian and American lecturers [F (1, 44) = 7.94, $p = 0.007$]. Comparing the means of both samples indicated that American lecturers evaluated this sub factor significantly higher than Australian lecturers. To continue, an ANOVA test showed that there was a significant difference in evaluation of this sub factor between Australian and American students [F (1, 148) = 39.03, $p = 0.00$]. Comparing the means of both samples indicated that American students evaluated this sub factor significantly higher than Australian students. The results showed that in both countries, students, lecturers and administrative staff believed that student centred interactivity was above average.

4.3.1.2 Socio-communication

Table 4.10 reports the means and standard deviations of the socio-communication sub-factor based on the academic position of Australian participants. As can be seen in this table, the highest mean regarding this sub factor belonged to administrative staff (*M* = 14.50, *SD* = 1.69). After administrative staff, the students reported this sub factor (*M* = 12.71, *SD* = 1.55) next highest and the lowest score was reported by the lecturers (*M* = 12.70, *SD* = 2.07). To investigate whether there are any differences in evaluation of socio- communication between students, lecturers and administrative staff, ANOVA was applied. The results showed that there was a significant effect of academic position on evaluation of student centred interactivity by participants of one faculty in an Australian university [F (2, 98) = 4.16, $p = 0.01$]. An LSD multiple comparison test between the three academic positions revealed that administrative staff reported this factor significantly higher than students and lecturers. However, the evaluation by lecturers and students of this sub-factor was the same. The results showed that all participants of one faculty in an Australian university believed socio- communication was above average.

In one faculty in a US university, as can be seen in this table, the highest mean of the answers to this sub factor belonged to students (*M* = 13.84, *SD* = 1.40). After students, the lecturers reported this factor (*M* = 13.06, *SD* = 1.06) next highest and the lowest score was reported by the administrative staff (*M* = 12.84, *SD* = 1.77). To investigate if there are any differences in evaluation of socio- communication between American students, lecturers and administrative staff, ANOVA was applied. The results showed that there was a significant main effect of academic position on evaluation of this sub factor by participants of one faculty in an American university [F (2, 115) = 6.21, $p = 0.003$]. An LSD multiple comparison test between the three levels of academic positions revealed that students reported this factor significantly higher than administrative staff and lecturers. However, the evaluation of this sub factor by lecturers and administrative staff was the same. The results showed that all participants of one faculty in a US university believed socio-communication to be above average.

Comparing the answers of participants of one faculty in an Australian university and one faculty in a US university showed that there were significant differences in evaluation of this sub factor between Australian and American administrative staff; Americans evaluated this sub factor higher than Australians [F (1, 20) = 4.46, $p = 0.04$]. Also the results of ANOVA revealed that there was no significant difference in evaluation of this sub factor between Australian and American lecturers

Table 4.10 Mean, SD, and F value of evaluation of socio-communication

Country	Students		Lecturers		Staff		*F*	*P*
	M	*SD*	*M*	*SD*	*M*	*SD*		
AUS participants	12.71	1.55	12.70	2.07	14.50	1.69	4.16	0.01*
USA participants	13.84	1.40	13.06	1.06	12.84	1.77	6.21	0.003**

* $p < 0.05$; ** $p < 0.01$

[F (1, 44) = 0.58, p = 0.45]. An ANOVA test showed that there was significant difference in evaluation of this sub factor between Australian and American students [F (1, 148) = 39.00, p = 0.00]. The results showed that in both countries, students, lecturers and administrative staff believed socio-communication to be above average.

4.3.1.3 Assessment

Table 4.11 reports the mean and standard deviation of the assessment sub factor based on the academic position of Australian and American participants. As can be seen in this table, the highest mean regarding this sub factor belonged to administrative staff (M = 11.87, SD = 1.35). After administrative staff, the students reported this sub factor (M = 10.18, SD = 1.21) as next highest and the lowest score was reported by the lecturers (M = 9.85, SD = 1.59). To investigate if there are any differences in evaluation of this sub factor between students, lecturers and administrative staff, ANOVA was applied. The results showed that there was a significant effect of academic position on evaluation of assessment on participants of one faculty in an Australian university [F (2, 98) = 7.21, p = 0.001]. An LSD multiple comparison test between the three academic positions revealed that administrative staff reported this factor significantly higher than students and lecturers. However, the evaluations by lecturers and students of this sub-factor were the same. The results showed that all participants of one faculty in an Australian university believed assessment to be above average.

In one faculty in a US university, as can be seen in this table, the highest mean of answers to this sub factor belonged to lecturers (M = 11.39, SD = 1.31). After them, students reported this factor (M = 10.59, SD = 1.11) next highest and the lowest score was reported by the administrative staff (M = 10.00, SD = 1.41). To investigate if there are any differences in evaluation of socio- communication between American students, lecturers and administrative staff, ANOVA was applied. The results showed that there was a significant main effect of academic position on evaluation of this sub factor by participants of one faculty in an American university [F (2, 115) = 6.78, p = 0.002]. An LSD multiple comparison test between the three levels of academic position revealed that lecturers reported this factor significantly higher than administrative staff and students. However, the evaluation of this sub factor by students and administrative staff was the same. The results showed that all participants of one faculty in a US university believed assessment was above average.

Table 4.11 Mean, SD, and F value of evaluation of assessment

Country	Students		Lecturers		Staff		*F*	*P*
	M	*SD*	*M*	*SD*	*M*	*SD*		
AUS participants	10.18	1.21	9.85	1.59	11.87	1.35	7.21	0.001**
USA participants	10.59	1.11	11.39	1.31	10.00	1.41	6.78	0.002**

** $p < 0.01$

Comparing the answers of participants of one faculty in an Australian university and one faculty in a US showed that there were significant differences in evaluation of this sub factor between Australian and American administrative staff; Australians evaluated this sub factor higher than Americans [F (1, 20) = 8.97, p = 0.007]. The results of ANOVA revealed that there were significant difference in evaluation of this sub factor between Australian and American lecturers [F (1, 44) = 12.69, p = 0.001] which indicated that Australian administrative staff evaluated the assessment sub factor more highly significantly than American administrative staff. An ANOVA test showed that there was a significant difference in evaluation of this sub factor between Australian and American students [F (1, 148) = 4.67, p = 0.03] illustrating that American students evaluated this sub factor significantly higher than Australian students. The results showed that in both countries, students, lecturers and administrative staff believed assessment was above average.

4.3.1.4 Learning Resources

Table 4.12 reports the means and standard deviations of the learning resources sub factor based on the academic position of Australian and American participants. As can be seen in this table, the highest mean regarding this sub factor belonged to lecturers (M = 4.37, SD = 0.67). After them, the administrative staff reported this sub factor (M = 4.12, SD = 0.64) next highest and the lowest score was reported by students (M = 3.54, SD = 0.71). To investigate if there are any differences in evaluation of this sub factor between students, lecturers and administrative staff, ANOVA was applied. The results showed that there was a significant main effect of academic position on evaluation of this sub factor by participants of one faculty in an Australian university [F (2, 98) = 11.41, p = 0.00]. An LSD multiple comparison test between the three academic positions revealed that students reported this factor significantly lower than administrative staff and lecturers. However, the evaluation by lecturers and administrative staff of this sub-factor was the same. The results showed that in Australia, students believed that learning resources were above average. However, lecturers and administrative staff assessed this sub factor as excellent.

In one faculty in a US university, as can be seen in this table, the highest mean of answers to this sub factor belonged to students (M = 4.37, SD = 0.58). Lecturers reported this factor next highest (M = 4.18, SD = 0.55) and the lowest score was reported by the administrative staff (M = 4.00, SD = 1.00). To investigate if there

Table 4.12 Mean, SD, and F value of evaluation of learning resources

Country	Students		Lecturers		Staff		*F*	*P*
	M	*SD*	*M*	*SD*	*M*	*SD*		
AUS participants	3.54	0.71	4.37	0.67	4.12	0.64	11.41	0.00***
USA participants	4.37	0.58	4.18	0.55	4.00	1.00	2.40	0.09

*** $p < 0.001$

are any differences in evaluation of this sub factor between American students, lecturers and administrative staff, ANOVA was applied. The results showed that there was no significant main effect of academic position on evaluation of this sub factor by 0.54% of participants of one faculty in an American university[F (2, 115) = 2.40, p = 0.09]. The results showed that in America, students, lecturers and administrative staff believed learning resources were available at an excellent level.

Comparing the answers of participants of one faculty in an Australian university and one faculty in a US university showed that there were no significant differences on evaluation of this sub factor between Australian and American administrative staff [F (1, 20) = 0.09, p = 0.75]. Also the results of ANOVA revealed that there was no significant difference in evaluation of this sub factor between Australian and American lecturers [F (1, 44) = 0.77, p = 0.38]. An ANOVA test showed that there was significant difference in evaluation of this sub factor between Australian and American students [F (1, 148) = 60.62, p = 0.00] illustrating that American students evaluated this sub factor significantly higher than Australian students. Comparing the results of participants of one faculty in an Australian university and participants of one faculty in an American university showed that only Australian students assessed this sub factor as above average while American students and lecturers and Australian lecturers and administrative staff believed learning resources were excellent.

4.3.1.5 Learning Environment

Table 4.13 reports the means and standard deviations of the learning environment sub factor based on the academic participants of one faculty in an Australian university and one faculty in a US university. As can be seen in this table, the highest mean of this sub factor in Australia belonged to administrative staff (M = 8.25, SD = 1.03). After them, the lecturers reported this sub factor (M = 7.75, SD = 1.06) as high and the lowest score was reported by students (M = 7.52, SD = 0.89). To investigate if there are any differences in evaluation of this sub factor between students, lecturers and administrative staff, ANOVA was applied. The results showed that there was no significant main effect of academic position on evaluation of this sub factor by participants of one faculty in an Australian university [F (2, 98) = 2.37, p = 0.09]. The results showed that in Australia, students and lecturers believed the learning environment was above average. Also administrative staff assessed this sub factor as at an excellent level.

Table 4.13 Mean, SD, and F value of evaluation of learning environment

Country	Students		Lecturers		Staff		*F*	*P*
	M	*SD*	*M*	*SD*	*M*	*SD*		
AUS participants	7.52	0.89	7.75	1.06	8.25	1.03	2.37	0.09
USA participants	8.04	0.86	7.70	0.73	7.53	0.51	3.28	0.04*

* $p < 0.05$

In one faculty in a US university, as can be seen in this table, the highest mean regarding this sub factor belonged to students ($M = 8.04$, $SD = 0.86$). Lecturers reported this factor next highest ($M = 7.70$, $SD = 0.73$) and the lowest score was reported by the administrative staff ($M = 7.53$, $SD = 0.51$). To investigate if there are any differences in evaluation of this sub factor between American students, lecturers and administrative staff, ANOVA was applied. The results showed that there was significant main effect of academic position on evaluation of this sub factor by participants of one faculty in a US university [F (2, 115) = 3.28, $p = 0.04$]. An LSD test showed that students evaluated this sub factor significantly higher than lecturers and administrative staff. There were no differences in evaluation by lecturers and administrative staff of this sub factor. The results showed that in America, students believed the learning environment was at an excellent level. However, lecturers and administrative staff assessed this sub factor as only above average.

Comparing the answers of participants of one faculty in an Australian university and one faculty in a US university showed that there were significant differences in evaluation of this sub factor between Australian and American administrative staff [F (1, 20) = 4.43, $p = 0.04$]; Australians significantly evaluated this sub factor higher than Americans. However, the results of ANOVA revealed that there was no significant difference in evaluation of this sub factor between Australian and American lecturers [F (1, 44) = 0.02, $p = 0.87$]. An ANOVA test showed that there was significant difference in evaluation of this sub factor between Australian and American students [F (1, 148) = 13.36, $p = 0.00$]; American students evaluated this sub factor significantly higher than Australian students. Comparing the results of Australians and Americans showed that Australian staff and American students assessed this sub factor at an excellent level while American lecturers and staff and Australian lecturers and students believed the learning environment was above average.

4.3.1.6 Pedagogical e-Practice

Table 4.14 reports the means and standard deviations of the pedagogical e-practice factor based on the academic positions of participants of one faculty in an Australian university and one faculty in a US university. As can be seen in this table, the highest mean of the pedagogical e-practice factor in Australia belonged to administrative staff ($M = 49.87$, $SD = 3.31$). After them, the lecturers reported the pedagogical e-practice factor ($M = 44.35$, $SD = 5.29$) as high and the lowest score was reported by students ($M = 43.04$, $SD = 4.01$). To investigate if there are any

Table 4.14 Mean, SD, and F value of evaluation of pedagogical e-practice

Country	Students		Lecturers		Staff		F	P
	M	SD	M	SD	M	SD		
AUS participants	43.04	4.01	44.35	5.29	49.87	3.31	9.43	0.00***
USA participants	47.26	3.11	47.45	2.60	45.61	2.81	1.91	0.15

*** $p < 0.001$

differences in evaluation of the pedagogical e-practice factor between students, lecturers and administrative staff, ANOVA was applied. The results showed that there was significant main effect of academic position on evaluation of the pedagogical e-practice factor by participants of one faculty in an Australian university [F (2, 98) = 9.43, p = 0.00]. An LSD test revealed that administrative staff evaluated this factor significantly higher than students and lecturers. However, there was no difference between the evaluation of students and lecturers of this factor. The results showed that all participants of one faculty in an Australian university believed pedagogical e-practice was above average.

In one faculty in a US university, as can be seen in this table, the highest mean of answers to the pedagogical e-practice factor belonged to lecturers (M = 47.45, SD = 2.60). After them, students reported this factor next highest (M = 47.26, SD = 3.11) and the lowest score was reported by the administrative staff (M = 45.61, SD = 2.81). To investigate if there are any differences in evaluation of the pedagogical e-practice factor between American students, lecturers and administrative staff, ANOVA was applied. The results showed that there was no significant main effect of academic position on evaluation of pedagogical e-practice factor by participants of one faculty in a US university [F (2, 115) = 1.91, p = 0.15]. The results showed that all participants of one faculty in a US university believed pedagogical e-practice was above average. Comparing the answers of participants of one faculty in an Australian university and one faculty in a US showed that there were significant differences in evaluation of the pedagogical e-practice factor between Australian and American administrative staff [F (1, 20) = 9.92, p = 0.005]; Australians significantly evaluated the pedagogical e-practice factor higher than Americans. However, the results of ANOVA revealed that there was significant difference in evaluation of the pedagogical e-practice factor between Australian and American lecturers [F (1, 44) = 6.63, p = 0.01]; Americans evaluated this factor significantly higher than Australians. To continue, an ANOVA test showed that there was significant difference in evaluation of the pedagogical e-practice factor between Australian and American students [F (1, 148) = 51.95, p = 0.00]; American students evaluated this factor significantly higher than Australian students. The results showed that in both countries, students, lecturers and administrative staff believed pedagogical e-practice to be above average (see Fig. 4.1).

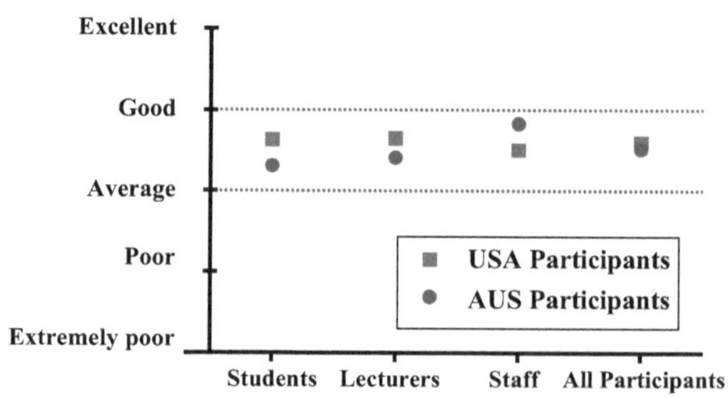

Fig. 4.1 Mean level of pedagogical e-practice

4.3.2 Technological E-Practice Factor Results

In this section, the participants' assessments of the technological e-practice factor and its sub factors, technological infrastructure, technological platforms, accessibility, reusability, and interface design are reported. The results of the Australian sample and the results of the American sample are presented first, followed by the comparative results between Australians and Americans.

4.3.2.1 Technological Infrastructure

Table 4.15 reports the means and standard deviations of the technological infrastructure sub factor based on the academic positions of the participants of one faculty in an Australian university and one faculty in a US university. As can be seen in this table, in Australia the highest mean of the technological infrastructure sub factor belonged to lecturers ($M = 7.75$, $SD = 1.44$). After them, the students reported this sub factor ($M = 7.66$, $SD = 0.97$) as high and the lowest score was reported by administrative staff ($M = 7.62$, $SD = 0.74$). To investigate if there are any differences in evaluation of the technological infrastructure sub factor between students, lecturers and administrative staff, ANOVA was applied. The results showed that there was no significant main effect of academic position on evaluation of the technological infrastructure sub factor by participants of one faculty in an Australian university [F (2, 98) = 0.06, $p = 0.93$] which means that there is no difference between the evaluation of lecturers, students and administrative staff of the technological infrastructure in Australia. The results showed that all participants of one faculty in an Australian university believed the technological infrastructure was above average.

In one faculty in a US university, as can be seen in this table, the highest mean regarding the technological infrastructure sub factor belonged to students ($M = 8.20$, $SD = 0.72$). After them lecturers reported this factor next highest ($M = 7.96$, $SD = 0.45$), and the lowest score was reported by the administrative staff ($M = 4.84$, $SD = 1.40$). Based on an ANOVA test, there was significant main effect of academic position on evaluation of technological infrastructure [F (2, 115) = 103.38, $p = 0.00$). An LSD multiple comparison test showed that students and lecturers evaluated this factor significantly higher than administrative staff. However, there was no significant difference between the evaluation of students and lecturers of this sub factor. The results of the technological infrastructure

Table 4.15 Mean, SD, and F value of evaluation of technological infrastructure

Country	Students		Lecturers		Staff		*F*	*P*
	M	*SD*	*M*	*SD*	*M*	*SD*		
AUS participants	7.66	0.97	7.75	1.44	7.62	0.74	0.06	0.93
USA participants	8.20	0.72	7.96	0.45	4.84	1.40	103.38	0.00***

*** $p < 0.001$

assessment showed that the staff believed this sub factor was at an average level, the lecturers believed this sub factor was above average and students believed it was at an excellent level.

Comparing the answers of participants of one faculty in an Australian university and one faculty in a US university showed that there were significant differences in evaluation of the technological infrastructure sub factor between Australian and American administrative staff [F (1, 20) = 26.35, $p = 0.00$]; Australians significantly evaluated the technological infrastructure sub factor higher than Americans. However, the results of ANOVA revealed that there was no significant difference in evaluation of this sub factor between Australian and American lecturers [F (1, 44) = 0.47, $p = 0.49$]. An ANOVA test showed that there was significant difference in evaluation between Australian and American students [F (1, 148) = 15.19, $p = 0.00$] illustrating that American students evaluated the technological infrastructure sub factor significantly higher than Australian students. The results showed that all participants of one faculty in an Australian university considered it above average however, participants of one faculty in a US university had different evaluations. American lecturers assessed this sub factor above average, students believed it to be excellent and staff believed technological infrastructure to be only average.

4.3.2.2 Technological Platform

Table 4.16 reports the means and standard deviations of the technological platform sub factor based on the academic participants of one faculty in an Australian university and one faculty in a US university. As can be seen in this table, in Australia the highest mean of the technological platform sub factor belonged to administrative staff ($M = 7.12$, $SD = 0.83$). After them, the lecturers reported the technological platform sub factor ($M = 6.75$, $SD = 1.33$) as high and the lowest score was reported by students ($M = 6.22$, $SD = 0.98$). To investigate if there are any differences in evaluation of the technological platform sub factor between students, lecturers and administrative staff, ANOVA was applied. It was found that there was significant main effect of academic position on evaluation of technological platforms [F (2, 98) = 3.95, $p = 0.02$). An LSD multiple comparison test showed that administrative staff evaluated technological platforms higher than lecturers and students. There was no significant difference in evaluation of this factor between students and lecturers. The results showed that all participants of one faculty in an Australian university believed the technological platform to be above average.

Table 4.16 Mean, SD, and F value of evaluation of technological platform

Country	Students		Lecturers		Staff		*F*	*P*
	M	*SD*	*M*	*SD*	*M*	*SD*		
AUS participants	6.22	0.98	6.75	1.33	7.12	0.83	3.95	0.02**
USA participants	6.95	0.88	7.60	0.69	8.30	0.75	17.66	0.00***

** $p < 0.01$; *** $p < 0.001$

In one faculty in a US university, as can be seen in this table, the highest mean regarding the technological platform sub factor belonged to administrative staff ($M = 8.30$, $SD = 0.75$). After them, lecturers reported this factor next highest ($M = 7.60$, $SD = 0.69$) and the lowest score was reported by the students ($M = 6.95$, $SD = 0.88$). To investigate if there are any differences in evaluation of the technological platform sub factor between American students, lecturers and administrative staff, ANOVA was applied. There was significant main effect of academic position on evaluation of technological platforms [F (2, 115) = 17.66, $p = 0.00$). An LSD multiple comparison test showed that administrative staff evaluated this sub factor significantly higher than lecturers and students. Also lecturers evaluated this sub factor significantly higher than students. The results of the technological platform assessment showed that the American staff believed it was excellent, however, the lecturers and students believed it was only above average.

Comparing the answers of participants of one faculty in an Australian university and one faculty in a US university showed that there were significant differences in evaluation of the technological platform sub factor between Australian and American administrative staff [F (1, 20) = 11.30, $p = 0.003$]; Americans significantly evaluated the technological platform sub factor higher than Australians. The results of ANOVA revealed that there was significant difference in evaluation of the technological platform sub factor between Australian and American lecturers [F (1, 44) = 7.80, $p = 0.008$]; Americans evaluated this sub factor significantly higher than Australians. To continue, an ANOVA test showed that there was significant difference in evaluation of the technological platform sub factor between Australian and American students [F (1, 148) = 22.44, $p = 0.00$]; American students evaluated this sub factor significantly higher than Australian students. The results of the technological platform assessment showed that all participants of one faculty in an Australian university and American lecturers and students had the same opinion namely above average however, American staff had a different assessment, namely excellent, for technological platform.

4.3.2.3 Accessibility

Table 4.17 reports the means and standard deviations of the accessibility sub factor based on the academic participants of one faculty in an Australian university and one faculty in a US university. As can be seen in this table, the highest mean

Table 4.17 Mean, SD, and F value of evaluation of accessibility

Country	Students		Lecturers		Staff		*F*	*P*
	M	*SD*	*M*	*SD*	*M*	*SD*		
AUS participants	3.84	0.55	3.80	0.69	4.12	0.64	0.93	0.39
USA participants	4.20	0.54	4.47	0.48	2.84	0.68	40.99	0.00***

*** $p < 0.001$

regarding the accessibility sub factor in Australia belonged to administrative staff ($M = 4.12$, $SD = 0.64$). After them, the students reported the accessibility sub factor ($M = 3.84$, $SD = 0.55$) as high and the lowest score was reported by lecturers ($M = 3.80$, $SD = 0.69$). To investigate if there are any differences in evaluation of the accessibility sub factor between students, lecturers and administrative staff, ANOVA was applied. The results showed that there was no significant main effect of academic position on evaluation of the accessibility sub factor by participants of one faculty in an Australian university [F (2, 98) = 0.93, $p = 0.39$]. The results showed that the Australian students and lecturers evaluated accessibility as above average whereas the staff evaluated this sub factor as excellent.

In one faculty in a US university, as can be seen in this table, the highest 0.68 mean of answers to the accessibility sub factor belonged to lecturers ($M = 4.47$, $SD = 0.48$). After them, students reported this factor next highest ($M = 4.20$, $SD = 0.54$) and the lowest score was reported by the administrative staff ($M = 2.84$, $SD = 0.68$). To investigate if there are any differences in evaluation of the accessibility sub factor between American students, lecturers and administrative staff, ANOVA was applied. The results showed that there was significant main effect of academic position on evaluation of the accessibility sub factor participants of one faculty in a USA university [F (2, 115) = 40.99, $p = 0.00$]. An LSD test showed that lecturers evaluated this sub factor significantly higher than students and administrative staff. Also, students evaluated this sub factor significantly higher than administrative staff. The results showed that American students and lecturers believed accessibility to be excellent. However, the staff evaluated this sub factor as only average.

Comparing the answers of participants of one faculty in an Australian university and one faculty in a US university showed that there were significant differences in evaluation of the accessibility sub factor between Australian and American administrative staff [F (1, 20) = 17.96, $p = 0.00$]; Australians significantly evaluated the accessibility sub factor higher than Americans. However, the results of ANOVA revealed that there was significant difference in evaluation of the accessibility sub factor between Australian and American lecturers [F (1, 44) = 14.51, $p = 0.00$]; Americans evaluated this sub factor significantly higher than Australians. An ANOVA test showed that there was significant difference in evaluation of the accessibility sub factor between Australian and American students [F (1, 148) = 16.09, $p = 0.00$] with American students evaluating accessibility significantly higher than Australian students. Comparing the results showed that American students and lecturers assessed accessibility higher than Australian students and lecturers. On the other hand Australian staff assessed this sub factor higher than American staff.

4.3.2.4 Reusability

Table 4.18 reports the means and standard deviations of the reusability sub factor based on the academic participants of one faculty in an Australian university and one faculty in a US university. As can be seen in this table, in Australia, the highest

Table 4.18 Mean, SD, and F value of evaluation of reusability

Country	Students		Lecturers		Staff		*F*	*P*
	M	*SD*	*M*	*SD*	*M*	*SD*		
AUS participants	3.77	0.56	4.10	0.96	3.87	0.35	1.93	0.15
USA participants	4.44	0.56	4.43	0.48	4.15	0.55	1.55	0.21

mean regarding the reusability sub factor belonged to lecturers ($M = 4.10$, $SD = 0.96$). After them, the administrative staff reported the sub factor ($M = 3.87$, $SD = 0.35$) as high and the lowest score was reported by students ($M = 3.77$, $SD = 0.56$). To investigate if there are any differences in evaluation of the reusability sub factor between students, lecturers and administrative staff, ANOVA was applied. The results showed that there was no significant main effect of academic position on evaluation of this sub factor by participants of one faculty in an Australian university [F (2, 98) = 1.93, $p = 0.15$]. The results showed that Australian students and staff believed reusability was above average. Also, the lecturers evaluated this sub factor at an excellent level.

In one faculty in a US university, as can be seen in this table, the highest mean of responses to the reusability sub factor belonged to students ($M = 4.44$, $SD = 0.56$). After them, lecturers reported this factor next highest ($M = 4.43$, $SD = 0.48$) and the lowest score was reported by the administrative staff ($M = 4.15$, $SD = 0.55$). To investigate if there are any differences in evaluation of the reusability sub factor between American students, lecturers and administrative staff, ANOVA was applied. The results showed that there was no significant main effect of academic position on evaluation of the reusability sub factor participants of one faculty in a USA university [F (2, 115) = 1.55, $p = 0.21$]. The results showed that all American participants believed reusability to be excellent.

Comparing the answers of participants of one faculty in an Australian university and one faculty in a US university showed that there were no significant differences in evaluation of the reusability sub factor between Australian and American administrative staff [F (1, 20) = 1.60, $p = 0.22$]. The results of ANOVA revealed that there was no significant difference in evaluation of this sub factor between Australian and American lecturers [F (1, 44) = 2.27, $p = 0.13$]. An ANOVA test showed that there was significant difference in evaluation of it between Australian and American students [F (1, 148) = 51.00, $p = 0.00$]; American students evaluated the reusability sub factor significantly higher than Australian students. Comparing the results showed that the American students and staff assessed this sub factor higher than Australian students and staff. On the other hand Australian staff assessed this sub factor higher than American staff.

4.3.2.5 Interface Design

Table 4.19 reports the means and standard deviations of the interface design sub factor based on the academic participants of one faculty in an Australian university

Table 4.19 Mean, SD, and F value of evaluation of interface design

Country	Students		Lecturers		Staff		*F*	*P*
	M	*SD*	*M*	*SD*	*M*	*SD*		
AUS participants	2.50	0.67	2.80	0.83	3.25	0.46	4.88	0.01*
USA participants	3.10	0.65	3.56	0.48	4.38	0.50	26.81	0.00***

* $p < 0.05$; *** $p < 0.001$

and one faculty in a US university. As can be seen in this table, the highest mean for the Australian sample belonged to administrative staff ($M = 3.25$, $SD = 0.46$). After them, the lecturers reported this sub factor ($M = 2.80$, $SD = 0.83$) as high and the lowest score was reported by students ($M = 2.50$, $SD = 0.67$). To investigate if there are any differences in evaluation of the interface design sub factor between students, lecturers and administrative staff, ANOVA was applied. The results showed that there was significant main effect of academic position on evaluation of the interface design sub factor by participants of one faculty in an Australian university [F (2, 98) = 4.88, $p = 0.01$]. An LSD test revealed that administrative staff evaluated this sub factor significantly higher than students. However, there were no significant differences between administrative staff and lecturers. The results showed that Australian students and lecturers believed interface design was at an average level. However, the staff believed in above average interface design.

In one faculty in a US university, as can be seen in this table, the highest mean of responses to the interface design sub factor belonged to administrative staff ($M = 4.38$, $SD = 0.50$). After them, lecturers reported this factor next highest ($M = 3.56$, $SD = 0.48$) and the lowest score was reported by the students ($M = 3.10$, $SD = 0.65$). To investigate if there are any differences in evaluation of the interface design sub factor between American students, lecturers and administrative staff, ANOVA was applied. The results showed that there was significant main effect of academic position on evaluation of this sub factor by participants of one faculty in a US university [F (2, 115) = 26.81, $p = 0.00$]. An LSD test showed that administrative staff evaluated the interface design sub factor significantly higher than lecturers and students. Also, lecturers evaluated this sub factor significantly higher than students. The results showed that American students and lecturers believed interface design was above average. However, the staff evaluated interface design as excellent.

Comparing the answers of participants of one faculty in an Australian university and one faculty in a US university showed that there were significant differences in evaluation of the interface design sub factor between Australian and American administrative staff [F (1, 20) = 26.46, $p = 0.00$]; Americans evaluated the interface design sub factor significantly higher than Australians. However, the results of ANOVA revealed that there was significant difference in evaluation of this sub factor between Australian and American lecturers [F (1, 44) = 14.99, $p = 0.00$]; American lecturers evaluated it significantly higher than Australian lecturers. To continue, an ANOVA test showed that there was significant difference in evaluation of the interface design sub factor between Australian and American students

[F (1, 148) = 29.84, p = 0.00] in that American students evaluated it significantly higher than Australian students. Overall, comparing the results showed that Australian students and lecturers believed interface design was average. However, the American students and lecturers believed that interface design was above average. On the other hand Australian staff placed interface design e-practice above average whereas the staff believed it to be excellent.

4.3.2.6 Technological e-Practice

Table 4.20 reports the means and standard deviations of the technological e-practice factor based on the academic participants of one faculty in an Australian university and one faculty in a US university. As can be seen in this table, in Australia, the highest mean of responses to the technological e-practice factor belonged to administrative staff (M = 26.00, SD = 1.85). After them, the lecturers reported this factor (M = 25.20, SD = 3.05) as high and the lowest score was reported by students (M = 24.01, SD = 2.63). To investigate if there are any differences in evaluation of the technological e-practice factor between students, lecturers and administrative staff, ANOVA was applied. The results showed that there was no significant main effect of academic position on evaluation of this factor by participants of one faculty in an Australian university [F (2, 98) = 3.06, p = 0.05]. The results showed that all participants of one faculty in an Australian university believed technological e-practice to be above average.

In one faculty in a US university, as can be seen in this table, the highest mean of answers to the technological e-practice factor belonged to lecturers (M = 28.04, SD = 1.24). After them, students reported this factor next highest (M = 26.90, SD = 1.71) and the lowest score was reported by the administrative staff (M = 24.53, SD = 1.39). To investigate if there are any differences in evaluation of technological factors between American students, lecturers and administrative staff, ANOVA was applied. The results showed that there was a significant main effect of academic position on evaluation of the technological factor by participants of one faculty in a US university [F (2, 115) = 20.78, p = 0.00]. An LSD test showed that lecturers evaluated the factor significantly higher than students and administrative staff. Also, students evaluated this factor significantly higher than administrative staff. The results showed that American students and staff believed technological e-practice to be above average. However, the lecturers believed in excellent technological e-practice.

Table 4.20 Mean, SD, and F value of evaluation of technological e-practice

Country	Students		Lecturers		Staff		*F*	*P*
	M	*SD*	*M*	*SD*	*M*	*SD*		
AUS participants	24.01	2.63	25.20	3.05	26.00	1.85	3.06	0.05
USA participants	26.90	1.71	28.04	1.24	24.53	1.39	20.78	0.00***

*** $p < 0.001$

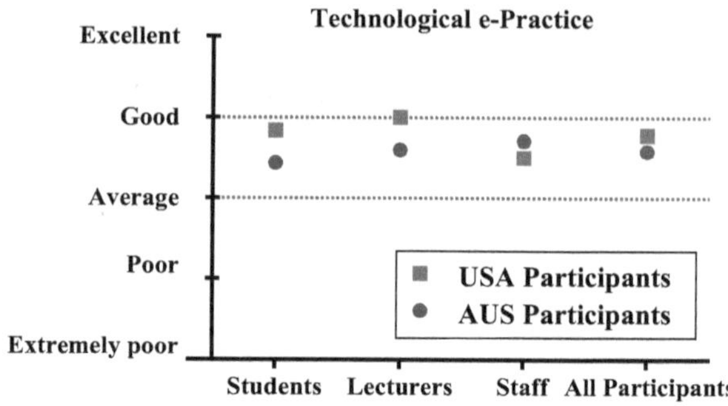

Fig. 4.2 Mean level of technological e-practice

Comparing the answers of participants of one faculty in an Australian university and one faculty in a US university showed that there were no significant differences in evaluation of the technological factor between Australian and American administrative staff [F (1, 20) = 2.25, p = 0.05]. However, the results of ANOVA revealed that there was significant difference in evaluation of this factor between Australian and American lecturers [F (1, 44) = 18.07, p = 0.00]; Americans evaluated this factor significantly higher than Australians. To continue, an ANOVA test showed that there was significant difference in evaluation of the technological factor between Australian and American students [F (1, 148) = 64.37, p = 0.00]; American students evaluated it significantly higher than Australian students. Overall, comparing the results showed that all participants of one faculty in an Australian university and American students as well as staff believed technological e-practice to be above average. However, the American lecturers believed that technological e-practice was excellent (Fig. 4.2).

4.3.3 *Instructional Design E-Practice Factor Results*

In this section, the responses to questions about the instructional design factor and its sub factors, namely clarifying expectation, personalization, learning scenarios, organizing resources and accuracy materials are reported. First the results of the Australian sample and then the results of the American sample are presented, followed by the comparative results between Australians and Americans.

4.3.3.1 Clarifying Expectation

Table 4.21 reports the means and standard deviations regarding the clarifying expectation sub factor based on the academic participants of one faculty in an Australian university and one faculty in a US university. As can be seen in this table, the highest mean, for Australia, belonged to administrative staff (M = 8.37, SD = 0.51). After them, the lecturers reported the clarifying expectation sub factor

Table 4.21 Mean, SD, and F value of evaluation of clarifying expectation

Country	Students		Lecturers		Staff		*F*	*P*
	M	*SD*	*M*	*SD*	*M*	*SD*		
AUS participants	7.69	0.99	7.85	1.08	8.37	0.51	1.79	0.17
USA participants	7.69	0.79	8.04	0.93	5.46	0.96	44.97	0.00***

*** $p < 0.001$

($M = 7.85$, $SD = 1.08$) as high and the lowest score was reported by students ($M = 7.69$, $SD = 0.99$). To investigate if there are any differences in evaluation of this sub factor between students, lecturers and administrative staff, ANOVA was applied. The results showed that there was no significant main effect of academic position on evaluation of the clarifying expectation sub factor by participants of one faculty in an Australian university [F (2, 98) = 1.79, $p = 0.17$]. The results showed that Australian students and lecturers believed clarifying expectation was above average. Also, the staff believed in excellent clarifying expectation.

In one faculty in a US university, as can be seen in this table, the highest mean of answers to the clarifying expectation sub factor belonged to lecturers ($M = 8.04$, $SD = 0.93$). After them, students reported this factor next highest ($M = 7.69$, $SD = 0.79$) and the lowest score was reported by the administrative staff ($M = 5.46$, $SD = 0.96$). To investigate if there are any differences in evaluation of the clarifying expectation sub factor between American students, lecturers and administrative staff, ANOVA was applied. The results showed that there was significant main effect of academic position on evaluation of the clarifying expectation sub factor by participants of one faculty in a US university [F (2, 115) = 44.97, $p = 0.00$]. An LSD test showed that administrative staff evaluated this sub factor significantly lower than students and lecturers. There were no differences between the evaluation of lecturers and students. The results showed that American staff believed clarifying expectation to be average. However, the students believed this sub factor was above average. On the other hand the lecturers believed in excellent clarifying expectation.

Comparing the answers of participants of one faculty in an Australian university and one faculty in a US university showed that there were significant differences in evaluation of the clarifying expectation sub factor between Australian and American administrative staff [F (1, 20) = 60.94, $p = 0.00$]; Australians significantly evaluated it higher than Americans. However, the results of ANOVA revealed that there was no significant difference in evaluation of this sub factor between Australian and American lecturers [F (1, 44) = 0.42, $p = 0.51$]. An ANOVA test showed that there was no significant difference in evaluation of the clarifying expectation sub factor between Australian and American students [F (1, 148) = 0.00, $p = 0.99$]. Overall, comparing the results showed that American and Australian students believed clarifying expectation to be above average. On the other hand, the American lecturers assessed this sub factor as excellent but Australian lecturers assessed clarifying expectation as above average. Surprisingly, the American staff gave an average assessment to clarifying expectation but Australian staff believed clarifying expectation was excellent.

4.3.3.2 Personalization

Table 4.22 reports the means and standard deviations of the personalization sub factor based on the academic participants of one faculty in an Australian university and one faculty in a US university. As can be seen in this table, the highest mean of the personalization sub factor belonged to administrative staff ($M = 3.37$, $SD = 0.51$). After them, the students reported the personalization sub factor ($M = 2.98$, $SD = 0.58$) as high and the lowest score was reported by lecturers ($M = 2.70$, $SD = 0.57$). To investigate if there are any differences in evaluation of the personalization sub factor between students, lecturers and administrative staff, ANOVA was applied. The results showed that there was significant main effect of academic position on evaluation of this sub factor by participants of one faculty in an Australian university [F (2, 98) = 4.29, $p = 0.01$]. An LSD test revealed that lecturers evaluated this sub factor significantly lower than administrative staff and students; however, there were no differences in evaluation of this sub factor between students and administrative staff. The results showed that Australian students and lecturers believed personalization to be at an average level. However, the staff believed that personalization was above average.

In one faculty in a US university, as can be seen in this table, the highest mean of responses to the personalization sub factor belonged to administrative staff ($M = 4.23$, $SD = 0.43$). After them, students reported this factor next highest ($M = 3.11$, $SD = 0.62$) and the lowest score was reported by the lecturers ($M = 2.44$, $SD = 0.69$). To investigate if there are any differences in evaluation of the sub factor between American students, lecturers and administrative staff, ANOVA was applied. The results showed that there was significant main effect of academic position on evaluation of the personalization sub factor by participants of one faculty in a US university [F (2, 115) = 34.98, $p = 0.00$]. An LSD test showed that administrative staff evaluated the personalization sub factor significantly higher than students and lecturers. Also, students evaluated this sub factor significantly higher than lecturers. The results showed that all participants of one faculty in a US university had different assessments. The students believed it to be above average, the lecturers believed average and the staff believed excellent.

Table 4.22 Mean, SD, and F value of evaluation of personalization

Country	Students		Lecturers		Staff		F	P
	M	SD	M	SD	M	SD		
AUS participants	2.98	0.58	2.70	0.57	3.37	0.51	4.29	0.01*
USA participants	3.11	0.62	2.44	0.69	4.23	0.43	34.98	0.00***

* $p < 0.05$; *** $p < 0.001$

Comparing the answers of participants of one faculty in an Australian university and one faculty in a US university showed that there were significant differences in evaluation of the personalization sub factor between Australian and American administrative staff [F (1, 20) = 16.47, p = 0.001]; Americans significantly evaluated it higher than Australians. However, the results of ANOVA revealed that there was no significant difference in evaluation of this sub factor between Australian and American lecturers [F (1, 44) = 1.67, p = 0.20]. An ANOVA test showed that there was no significant difference in evaluation of the personalization sub factor between Australian and American students [F (1, 148) = 0.00, p = 0.99]. Overall, comparing the results showed that Australian and American lecturers believed personalization practice to be average. The American students assessed it as above average but Australian students placed it at an average level. On the other hand, American staff assessed this sub factor as excellent while Australian staff assessed it above average.

4.3.3.3 Learning Scenarios

Table 4.23 reports the means and standard deviations regarding the learning scenarios sub factor based on the academic participants of one faculty in an Australian university and one faculty in a US university. As can be seen in this table, the highest mean of the learning scenarios sub factor belonged to administrative staff (M = 7.12, SD = 1.12). After them, the lecturers reported this sub factor (M = 6.95, SD = 1.35) as high and the lowest score was reported by students (M = 6.63, SD = 1.04). To investigate if there are any differences in evaluation of the learning scenarios sub factor between students, lecturers and administrative staff, ANOVA was applied. The results showed that there was no significant main effect of academic position on evaluation of the learning scenarios sub factor by participants of one faculty in an Australian university [F (2, 98) = 1.14, p = 0.32]. The results showed that all participants of one faculty in an Australian university believed learning scenarios to be above average.

In one faculty in a US university, as can be seen in this table, the highest mean of responses to the learning scenarios sub factor belonged to lecturers (M = 8.00, SD = 0.91). After them, students reported this factor next highest (M = 7.18, SD = 0.98) and the lowest score was reported by the administrative staff (M = 6.61, SD = 0.65). To investigate if there are any differences in evaluation of this sub factor between American students, lecturers and administrative staff, ANOVA was

Table 4.23 Mean, SD, and F value of evaluation of learning scenarios

Country	Students		Lecturers		Staff		F	P
	M	SD	M	SD	M	SD		
AUS participants	6.63	1.04	6.95	1.35	7.12	1.12	1.14	0.32
USA participants	7.18	0.98	8.00	0.91	6.61	0.65	10.97	0.00***

*** $p < 0.001$

applied. The results showed that there was significant main effect of academic position on evaluation of learning scenarios sub factor by participants of one faculty in a US university [F (2, 115) = 10.97, p = 0.00]. An LSD test showed that lecturers evaluated the learning scenarios sub factor significantly higher than students and administrative staff. Also, students evaluated this sub factor significantly higher than administrative staff. The results showed that American students and staff believed learning scenarios to be above average. However, the lecturers believed learning scenarios were excellent.

Comparing the answers of participants of one faculty in an Australian university and one faculty in a US university showed that there were no significant differences in evaluation of the learning scenarios sub factor between Australian and American administrative staff [F (1, 20) = 1.75, p = 0.20]. However, the results of ANOVA revealed that there was significant difference in evaluation of it between Australian and American lecturers [$F\,(1, 44) = 9.58, p = 0.003$]; Americans evaluated this sub factor significantly higher than Australians. An ANOVA test showed significant difference in evaluation of the learning scenarios sub factor between Australian and American students [F (1, 148) = 11.09, p = 0.001]; American students evaluated this sub factor significantly higher than Australian students. Overall, comparing the results showed that all participants of one faculty in an Australian university and American students and staff believed learning scenarios practice was above average. However, the American lecturers believed learning scenarios practice was at an excellent level.

4.3.3.4 Organizing Resources

Table 4.24 reports the means and standard deviations of the organizing resources sub factor based on the academic participants of one faculty in an Australian university and one faculty in a US university. As can be seen in this table in Australia, the highest mean of the organizing resources sub factor belonged to administrative staff (M = 3.87, SD = 0.35). After them, the lecturers reported this sub factor (M = 3.25, SD = 0.85) as high and the lowest score was reported by students (M = 3. 22, SD = 0.65). To investigate if there are any differences in evaluation of the organizing resources sub factor between students, lecturers and administrative staff, ANOVA was applied. The results showed that there was significant main effect of academic position on evaluation of the organizing resources sub factor by participants of one faculty in an Australian university

Table 4.24 Mean, SD, and F value of evaluation of organizing resources

Country	Students		Lecturers		Staff		*F*	*P*
	M	*SD*	*M*	*SD*	*M*	*SD*		
AUS participants	3.22	0.65	3.25	0.85	3.87	0.35	3.26	0.094
USA participants	3.39	0.62	3.45	0.46	4.61	0.50	24.76	0.00***

*** $p < 0.001$

[F (2, 98) = 3.26, p = 0.094]. An LSD test showed that administrative staff evaluated this sub factor significantly higher than lecturers and students. However, there were no differences in evaluation of this sub factor between students and lecturers. The results showed that all participants of one faculty in an Australian university had the same assessment namely that organizing resources were poor.

In one faculty in a US university, as can be seen in this table, the highest mean of responses to the organizing resources sub factor belonged to administrative staff (M = 4.61, SD = 0.50). After them, lecturers reported this factor next highest (M = 3.45, SD = 0.46) and the lowest score was reported by the students (M = 3.39, SD = 0.62). To investigate if there are any differences in evaluation of the organizing resources sub factor between American students, lecturers and administrative staff, ANOVA was applied. The results showed that there was significant main effect of academic position on evaluation of the organizing resources sub factor participants of one faculty in a USA university [F (2, 115) = 24.76, p = 0.00]. An LSD test showed that administrative staff evaluated it significantly higher than lecturers and students. There were no differences between the evaluation of lecturers and students. The results showed that American students and lecturers believed organizing resources practice was poor. However, the staff believed organizing resources practice was at an average level.

Comparing the answers of participants of one faculty in an Australian university and one faculty in a US university showed that there were significant differences in evaluation of the organizing resources sub factor between Australian and American administrative staff [F (1, 20) = 13.05, p = 0.002]; Americans significantly evaluated this sub factor higher than Australians. However, the results of ANOVA revealed that there was no significant difference in evaluation of the organizing resources sub factor between Australian and American lecturers [F (1, 44) = 1.04, p = 0.31]. Also, an ANOVA test showed that there was no significant difference in evaluation of it between Australian and American students [F (1, 148) = 2.60, p = 0.10]. Overall, comparing the results showed that all participants of one faculty in an Australian university and American students and lecturers believed organizing resources practice was at a poor level. However, the American staff believed organizing resources practice to be average.

4.3.3.5 Accuracy Materials

Table 4.25 reports the means and standard deviations of the accuracy materials sub factor based on the academic participants of one faculty in an Australian university

Table 4.25 Mean, SD, and F value of evaluation of accuracy materials

Country	Students		Lecturers		Staff		*F*	*P*
	M	*SD*	*M*	*SD*	*M*	*SD*		
AUS participants	6.14	0.94	5.75	1.33	7.00	0.92	4.19	0.01*
USA participants	6.20	0.76	5.11	1.19	8.46	0.96	60.24	0.00***

* $p < 0.05$; *** $p < 0.001$

and one faculty in a US university. As can be seen in this table, in Australia, the highest mean of accuracy materials belonged to administrative staff ($M = 7.00$, $SD = 0.92$). After them, the students reported accuracy materials ($M = 6.14$, $SD = 0.94$) as high and the lowest score was reported by lecturers ($M = 5.75$, $SD = 1.33$). To investigate if there are any differences in evaluation of accuracy materials between students, lecturers and administrative staff, ANOVA was applied. The results showed that there was significant main effect of academic position on evaluation of accuracy materials by participants of one faculty in an Australian university [F (2, 98) = 4.19, $p = 0.01$]. An LSD test illustrated that administrative staff evaluated this sub factor significantly higher than students and lecturers; however, there were no significant differences on evaluation of this sub factor between students and lecturers. The results showed that Australian students and staff believed that accuracy materials are at an above average level. However, the lecturers believed accuracy materials were only average.

In one faculty in a US university, as can be seen in this table, the highest mean of answers regarding the accuracy materials belonged to administrative staff ($M = 8.46$, $SD = 0.96$). Next highest came students' reports of this factor ($M = 6.20$, $SD = 0.76$) and the lowest score was reported by the lecturers ($M = 5.11$, $SD = 1.19$). To investigate if there are any differences in evaluation of accuracy materials between American students, lecturers and administrative staff, ANOVA was applied. The results showed that there was significant main effect of academic position on evaluation of accuracy materials by participants of one faculty in a US university [F (2, 115) = 60.24, $p = 0.00$]. An LSD test showed that administrative staff evaluated accuracy materials significantly higher than lecturers and students. Also, students evaluated this sub factor significantly higher than lecturers. The results showed that American students believed this sub factor was at an above average level, the lecturers believed it was at an average level and the staff believed accuracy materials practice was excellent. Comparing the answers of participants of one faculty in an Australian university and one faculty in a US university showed that there were significant differences in evaluation of accuracy materials between Australian and American administrative staff [F (1, 20) = 11.66, $p = 0.003$]; Americans significantly evaluated accuracy materials higher than Australians. However, the results of ANOVA revealed that there was no significant difference in evaluation of accuracy materials between Australian and American lecturers [F (1, 44) = 2.85, $p = 0.09$]. To continue, an ANOVA test showed that there was no significant difference in evaluation of accuracy materials between Australian and American students [F (1, 148) = 0.21, $p = 0.64$]. Overall, comparing the results showed that Australian students and lecturers as well as American students and lecturers had the same perspective in that they believed accuracy materials practice to be above average. The Australian staff believed accuracy materials practice was above average however, the American staff believed in an excellent level of accuracy materials.

4.3.3.6 Instructional Design e-Practice

Table 4.26 reports the means and standard deviations of the instructional design practice factor based on the academic participants of one faculty in an Australian university and one faculty in a US university. As can be seen in this table, the highest mean of instructional design practice belonged to administrative staff (*M* = 29.75, *SD* = 1.83). After them, the students reported instructional design practice (*M* = 26.67, *SD* = 2.50) as high and the lowest score was reported by lecturers (*M* = 26.50, *SD* = 2.60). To investigate if there are any differences in evaluation of instructional design practice between students, lecturers and administrative staff, ANOVA was applied. The results showed that there was significant main effect of academic position on evaluation of instructional design practice participants of one faculty in an Australian university [F (2, 98) = 5.83, p = 0.004]. An LSD test showed that administrative staff evaluated this factor significantly higher than students and lecturers but there were no significant differences in evaluation of this factor between students and lecturers. Overall, regarding the level of practice assessment it seems that all participants of one faculty in an Australian university had the same assessment placing it in an above average level.

In one faculty in a US university, as can be seen in this table, the highest mean of responses to instructional design practice belonged to administrative staff (*M* = 29.38, *SD* = 1.75). After them, students reported this factor next highest (M = 27.59, SD = 2.23) and the lowest score was reported by the lecturers (M = 27.06, SD = 2.21). To investigate if there are any differences in evaluation of instructional design practice between American students, lecturers and administrative staff, ANOVA was applied. The results showed that there was significant main effect of academic position on evaluation of instructional design practice by participants of one faculty in a US university [F (2, 115) = 5.01, p = 0.008]. An LSD test showed that administrative staff evaluated this factor significantly higher than students and lecturers but there were no significant differences in evaluation of this factor between students and lecturers. The results showed that all participants of one faculty in a US university gave it the same assessment of above average.

Comparing the answers of participants of one faculty in an Australian university and one faculty in a US university showed that there was no significant differences in evaluation of instructional design practice between Australian and American administrative staff [F (1, 20) = 0.20, p = 0.65]. Furthermore, the results of ANOVA revealed that there was no significant difference in evaluation of

Table 4.26 Mean, SD, and F value of evaluation of instructional design e-practice

Country	Students		Lecturers		Staff		*F*	*P*
	M	*SD*	*M*	*SD*	*M*	*SD*		
AUS participants	26.67	2.50	26.50	2.60	29.75	1.83	5.83	0.004**
USA participants	27.59	2.23	27.06	2.21	29.38	1.75	5.01	0.008**

** $p < 0.01$

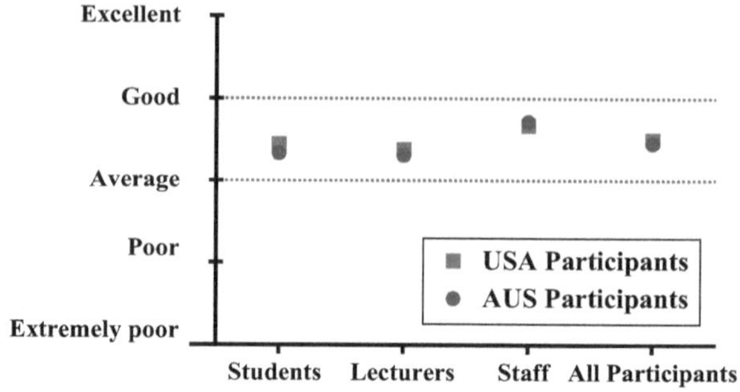

Fig. 4.3 Mean level of instructional design e-practice

instructional design practice between Australian and American lecturers [F (1, 44) = 0.61, p = 0.43]. However, ANOVA test showed that there was significant difference in evaluation of instructional design practice between Australian and American students [F (1, 148) = 5.61, p = 0.01]; American students evaluated instructional design practice significantly higher than Australian students. All participants of one faculty in an Australian and one faculty in a US university gave the same assessment namely that instructional design practice was above average (see Fig. 4.3).

4.3.4 *Organizational E-Practice Factor Results*

In this section, the participants' assessments of the organizational factor and its sub factors namely institutional affairs, administrative affairs, research development and precedent reputation are reported. The Australian results and then the results of the Americans are reported followed by the comparative results of the Australians and Americans.

4.3.4.1 Institutional Affairs

Table 4.27 reports the means and standard deviations of the institutional affairs sub factor based on the academic participants of one faculty in an Australian university and one faculty in a US university. As can be seen in this table, in Australia the highest mean of the institutional affairs sub factor belonged to administrative staff (M = 3.84, SD = 0.35). After them, the students reported the institutional affairs

Table 4.27 Mean, SD, and F value of evaluation of institutional affairs

Country	Students		Lecturers		Staff		F	P
	M	SD	M	SD	M	SD		
AUS participants	3.50	0.58	3.40	0.68	3.84	0.35	1.87	0.15
USA participants	3.85	0.55	3.65	0.62	4.07	0.27	2.66	0.07

sub factor ($M = 3.50$, $SD = 0.58$) as high and the lowest score was reported by lecturers ($M = 3.40$, $SD = 0.68$). To investigate if there are any differences in evaluation of the institutional affairs sub factor between students, lecturers and administrative staff, ANOVA was applied. The results showed that there was no significant main effect of academic position on evaluation of the institutional affairs sub factor by participants of one faculty in an Australian university [F (2, 98) = 1.87, $p = 0.15$]. The results showed that all participants of one faculty in an Australian university believed institutional affairs to be above average.

In one faculty in a US university, as can be seen in this table, the highest mean of answers to the institutional affairs sub factor belonged to administrative staff ($M = 4.07$, $SD = 0.\ 27$). After them, students reported this factor next highest ($M = 3.85$, $SD = 0.55$) and the lowest score was reported by the lecturers ($M = 3.65$, $SD = 0.62$). To investigate if there are any differences in evaluation of this sub factor between American students, lecturers and administrative staff, ANOVA was applied. The results showed that there was no significant main effect of academic position on evaluation of the institutional affairs sub factor by participants of one faculty in a US university [F (2, 115) = 2.66, $p = 0.07$]. The results showed that American students and lecturers believed institutional affairs to be above average. However, the staff believed institutional affairs practice was excellent.

Comparing the answers of participants of one faculty in an Australian university and one faculty in a US university showed that there were no significant differences in evaluation of the institutional affairs sub factor between Australian and American administrative staff [F (1, 20) = 2.13, $p = 0.16$]. Furthermore, the results of ANOVA revealed that there was no significant difference in evaluation of the institutional affairs sub factor between Australian and American lecturers [F (1, 44) = 1. 77, $p = 0.19$]. To continue, an ANOVA test showed that there was significant difference in evaluation of this sub factor between Australian and American students [F (1, 148) = 14.33, $p = 0.00$] illustrating that American students evaluated the institutional affairs sub factor significantly higher than Australian students. Overall, comparing the results showed that all participants of one faculty in an Australian university together with American students and lecturers had the same perspective believing institutional affairs practice to be above average. However, the American staff believed institutional affairs practice was at an excellent level.

4.3.4.2 Administrative Affairs

Table 4.28 reports the means and standard deviations of the administrative affairs sub factor based on the academic participants of one faculty in an Australian university and one faculty in a US university. As can be seen in this table, the highest mean of administrative affairs belonged to administrative staff ($M = 4.12$, $SD = 0.35$). After them, the lecturers reported administrative affairs ($M = 3.55$, $SD = 0.75$) as high and the lowest score was reported by students ($M = 3.33$, $SD = 0.65$). To investigate if there are any differences in evaluation of

Table 4.28 Mean, SD, and F value of evaluation of administrative affaires

Country	Students		Lecturers		Staff		*F*	*P*
	M	*SD*	*M*	*SD*	*M*	*SD*		
AUS participants	3.33	0.65	3.55	0.75	4.12	0.35	5.44	0.006**
USA participants	3.25	0.46	4.04	0.61	4.38	0.50	43.03	0.00***

** $p < 0.01$; *** $p < 0.001$

administrative affairs between students, lecturers and administrative staff, ANOVA was applied. The results showed that there was significant main effect of academic position on evaluation of administrative affairs by participants of one faculty in an Australian university [F (2, 98) = 5.44, p = 0.006]. LSD test revealed that administrative staff evaluated this sub factor significantly higher than students and lecturers; however, there were no significant differences in evaluation of this sub factor between students and lecturers. The results showed that Australian students and lecturers believed administrative affairs to be above average. However, the Australian staff believed administrative affairs were excellent.

In one faculty in a US university, as can be seen in this table, the highest mean of responses to administrative affairs belonged to administrative staff (M = 4.38, SD = 0.50). After them, lecturers reported this factor next highest (M = 4.04, SD = 0.61) and the lowest score was reported by the students (M = 3.25, SD = 0.46). To investigate if there are any differences in evaluation of administrative affairs between American students, lecturers and administrative staff, ANOVA was applied. The results showed that there was significant main effect of academic position on evaluation of administrative affairs by participants of one faculty in a US university [F (2, 115) = 43.03, p = 0.00]. An LSD test showed that students evaluated administrative affairs significantly lower than lecturers and administrative staff. There were no differences in evaluation of lecturers and administrative staff regarding administrative affairs. The results showed that American staff and lecturers believed administrative affairs were at an excellent level. However, the American students believed administrative affairs were only above average.

Comparing the answers of participants of one faculty in an Australian university and one faculty in a US university showed that there were no significant differences in evaluation of administrative affairs between Australian and American administrative staff [F (1, 20) = 1.60, p = 0.22]. Moreover, the results of ANOVA revealed that there was no significant difference in evaluation of administrative affairs between Australian and American lecturers [F (1, 44) = 5.99, p = 0.01] although Americans evaluated this sub factor higher than Australians. To continue, ANOVA test showed that there was no significant difference in evaluation between Australian and American students [F (1, 148) = 0.77, p = 0.37]. Overall, comparing the results showed that Australian students and lecturers as well as American students believed administrative affairs were above average. However, Australian staff and American staff and lecturers believed administrative affairs were excellent.

4.3.4.3 Research Development

Table 4.29 reports the means and standard deviations of the research development sub factor based on the academic participants of one faculty in an Australian university and one faculty in a US university. As can be seen in this table, the highest mean of the research development sub factor belonged to students (M = 4. 30, SD = 0.62). After them, the administrative staff reported it (M = 4.00, SD = 0.75) as high and the lowest score was reported by lecturers (M = 3.60, SD = 0.75). To investigate if there are any differences in evaluation of the research development sub factor between students, lecturers and administrative staff, ANOVA was applied. The results showed that there was significant main effect of academic position on evaluation of the research development sub factor by participants of one faculty in an Australian university [F (2, 98) = 9.18, p = 0.00]. LSD test showed that lecturers evaluated this sub factor significantly lower than students and administrative staff. However, there were no significant differences in evaluation of this sub factor between students and administrative staff. The results showed that Australian students and staff believed research development was excellent. However, the Australian lecturers believed research development was only above average.

In one faculty in a US university, as can be seen in this table, the highest mean of answers to the research development sub factor belonged to administrative staff (M = 4.15, SD = 1.14). After them, lecturers reported this factor next highest (M = 3.91, SD = 0.49) and the lowest score was reported by the students (M = 3.80, SD = 0.77). To investigate if there are any differences in evaluation of the research development sub factor between American students, lecturers and administrative staff, ANOVA was applied. The results showed that there was no significant main effect of academic position on evaluation of this sub factor by participants of one faculty in a US university [F (2, 115) = 1.17, p = 0.31]. The results showed that American students and lecturers believed research development was above average while the American staff assessed research development as excellent.

Comparing the answers of participants of one faculty in an Australian university and one faculty in a US university showed that there were no significant differences in evaluation of the research development sub factor between Australian and American administrative staff [F (1, 20) = 0. 11, p = 0.74]. However, the results of ANOVA revealed that there was significant difference in evaluation of this sub factor between Australian and American lecturers [F (1, 44) = 5.99, p = 0.01];

Table 4.29 Mean, SD, and F value of evaluation of research development

Country	Students		Lecturers		Staff		*F*	*P*
	M	*SD*	*M*	*SD*	*M*	*SD*		
AUS participants	4.30	0.62	3.60	0.75	4.00	0.75	9.18	0.00***
USA participants	3.80	0.77	3.91	0.49	4.15	1.14	1.17	0.31

*** $p < 0.001$

Americans evaluated this sub factor significantly higher. To continue, ANOVA test showed that there was significant difference in evaluation of the sub factor between Australian and American students [F (1, 148) = 0.77, p = 0.37] illustrating that American students evaluated it significantly higher than Australian students. Regarding the level of practice assessment, Australian lecturers and American lecturers and students believed research development practice to be above average, whereas Australian staff and students as well as American staff believed research development practice was excellent.

4.3.4.4 Precedent Reputation

Table 4.30 reports the means and standard deviations of the precedent reputation sub factor based on the academic participants of one faculty in an Australian university and one faculty in a US university. As can be seen in this table, the highest mean of the precedent reputation sub factor in the Australian sample belonged to administrative staff (M = 4.12, SD = 0.35). After them, the students reported the sub factor (M = 3.28, SD = 0.58) as high and the lowest score was reported by lecturers (M = 3.00, SD = 0.72). To investigate if there are any differences in evaluation of the precedent reputation sub factor between students, lecturers and administrative staff, ANOVA was applied. The results showed that there was significant main effect of academic position on evaluation of the precedent reputation sub factor on participants of one faculty in an Australian university [F (2, 98) = 9.89, p = 0.00]. LSD test revealed that administrative staff evaluated this sub factor significantly higher than students and lecturers. There were no significant differences between evaluation of students and lecturers regarding this sub factor. The results showed that Australian students and lecturers believed precedent reputation was above average. However, Australian staff believed precedent reputation was at an excellent level.

In one faculty in a US university, as can be seen in this table, the highest mean of responses to the precedent reputation sub factor belonged to administrative staff (M = 4.30, SD = 0.63). After them, lecturers reported this factor next highest (M = 3.96, SD = 0.61) and the lowest score was reported by the students (M = 3.52, SD = 0. 61). To investigate if there are any differences in evaluation of the precedent reputation sub factor between American students, lecturers and administrative staff, ANOVA was applied. The results showed that there was significant main effect of academic position on evaluation of this sub factor by

Table 4.30 Mean, SD, and F value of evaluation of precedent reputation

Country	Students		Lecturers		Staff		F	P
	M	SD	M	SD	M	SD		
AUS participants	3.28	0.58	3.00	0.72	4.12	0.35	9.89	0.00***
USA participants	3.52	0.61	3.96	0.61	4.30	0.63	11.70	0.00***

*** $p < 0.001$

participants of one faculty in a US university [F (2, 115) = 11.70, p = 0.00]. LSD test showed that students evaluated the sub factor significantly lower than lecturers and administrative staff. There were no differences between evaluation by lecturers and administrative staff of the precedent reputation sub factor. American students and lecturers believed precedent reputation to be above average. On the other hand, the American staff assessed precedent reputation as excellent.

Comparing the answers of participants of one faculty in an Australian university and one faculty in a US university showed that there were no significant differences in evaluation of the precedent reputation sub factor between Australian and American administrative staff [F (1, 20) = 0.55, p = 0.46]. However, the results of ANOVA revealed that there was significant difference in evaluation of precedent reputation sub factor between Australian and American lecturers [F (1, 44) = 23.22, p = 0.00]; Americans evaluated this sub factor significantly higher than Australians. Moreover, ANOVA test showed that there was significant difference in evaluation of this sub factor between Australian and American students [F (1, 148) = 6.13, p = 0.01] illustrating that American students evaluated the precedent reputation sub factor significantly higher than Australian students. Australian and American students and lecturers believed that precedent reputation was above average. However, Australian and American staff believed precedent reputation was at an excellent level.

4.3.4.5 Organizational e-Practice

Table 4.31 reports the means and standard deviations of the organizational practice factor based on the participants of one faculty in an Australian university and one faculty in a US university. As can be seen in this table, the highest mean of the organizational practice factor in the Australian sample belonged to administrative staff (M = 16.12, SD = 0.83). After them, the students reported the organizational practice factor (M = 14.43, SD = 1.62) as high and the lowest score was reported by lecturers (M = 13.55, SD = 1.63). To investigate if there are any differences in evaluation of this factor between students, lecturers and administrative staff, ANOVA was applied. The results showed that there was significant main effect of academic position on evaluation of organizational practice factor by participants of one faculty in an Australian university[F (2, 98) = 7.62, p = 0.001]. LSD test showed that administrative staff evaluated this factor significantly higher than

Table 4.31 Mean, SD, and F value of evaluation of organizational e-practice

Country	Students		Lecturers		Staff		*F*	*P*
	M	*SD*	*M*	*SD*	*M*	*SD*		
AUS participants	14.43	1.62	13.55	1.63	16.12	0.83	7.62	0.001**
USA participants	14.44	1.42	15.58	1.57	16.92	1.65	18.24	0.00***

** $p < 0.01$; *** $p < 0.001$

students and lecturers. Also, students evaluated this factor significantly higher than lecturers. Overall, the results showed that Australian students and lecturers believed organizational practice to be above average. However, the Australian staff believed organizational practice was excellent.

In one faculty in a US university, as can be seen in this table, the highest mean of responses to the organizational practice factor belonged to administrative staff ($M = 16.92$, $SD = 1.65$). After them, lecturers reported this factor next highest ($M = 15.58$, $SD = 1.57$) and the lowest score was reported by the students ($M = 14.44$, $SD = 1.42$). To investigate if there are any differences in evaluation of this factor between American students, lecturers and administrative staff, ANOVA was applied. The results showed that there was significant main effect of academic position on evaluation of the organizational practice factor by participants of one faculty in a US university [F (2, 115) = 18.24, $p = 0.00$]. LSD test showed that administrative staff evaluated the factor significantly higher than lecturers and students. Also, lecturers evaluated the factor significantly higher than students. American students and lecturers believed organizational practice was above average. On the other hand the American staff assessed organizational practice as excellent.

Comparing the answers of participants of one faculty in an Australian university and one faculty in a US university showed that there were no significant differences in evaluation of the organizational practice factor between Australian and American administrative staff [F (1, 20) = 1.58, $p = .22$]. However, the results of ANOVA revealed that there was significant difference in evaluation of this factor between Australian and American lecturers [F (1, 44) = 17.89, $p = 0.00$]; American lecturers evaluated this factor significantly higher than Australians. To continue, ANOVA test showed that there was no significant difference in evaluation of the organizational practice factor between Australian and American students [F (1, 148) = 0.003, $p = 0.95$]. Australian and American students and lecturers believed organizational practice to be above average. However, Australian and American staff believed organizational practice was excellent (see Fig. 4.4).

Fig. 4.4 Mean level of organizational e-practice

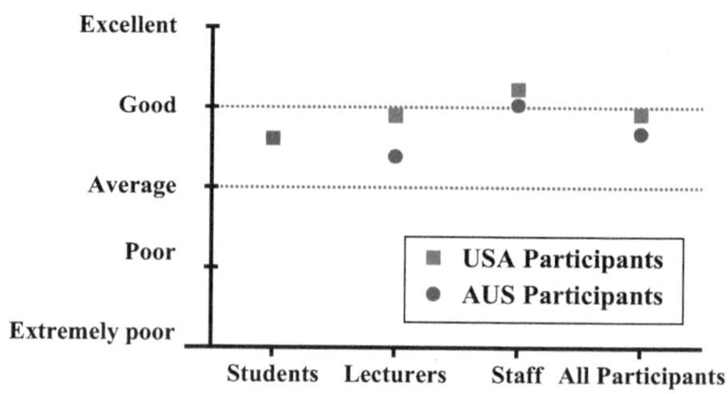

4.3.5 Support E-Practice Factor Results

This factor was evaluated by three sub factor namely administrative support, technical support and academic support. The results based on participants' assessment are reported in this section. First the result of the Australian sample and then the result of the American sample are presented, followed by the comparative results of Australians and Americans.

4.3.5.1 Administrative Support

Table 4.32 reports the means and standard deviations of the administrative support sub factor based on the academic participants of one faculty in an Australian university and one faculty in a US university. As can be seen in this table, in Australia the highest mean of the administrative support sub factor belonged to administrative staff ($M = 4.00$, $SD = 0.53$). After them, the students reported the administrative support sub factor ($M = 3.46$, $SD = 0.50$) as high and the lowest score was reported by lecturers ($M = 3.30$, $SD = 0.47$). To investigate if there are any differences on evaluation of this sub factor between students, lecturers and administrative staff, ANOVA was applied. The results showed that there was no significant main effect of academic position on evaluation of the administrative support sub factor on participants of one faculty in an Australian university [F (2, 98) = 5.68, $p = 0.005$]. LSD test showed that administrative staff evaluated this sub factor significantly higher than students and lecturers. However, students and lecturers evaluated this sub factor the same. The results showed that Australian students and lecturers believed administrative support was above average. However, Australian staff believed administrative support was excellent.

In one faculty in a US university, as can be seen in this table, the highest mean of responses to the administrative support sub factor belonged to administrative staff ($M = 4.15$, $SD = 0.37$). After them, students assessed this factor next highest ($M = 3.55$, $SD = 0.57$) and the lowest score was reported by the lecturers ($M = 3.39$, $SD = 0.47$). To investigate if there are any differences in evaluation of this sub factor between American students, lecturers and administrative staff, ANOVA was applied. The results showed that there was significant main effect of academic position on evaluation of the administrative support sub factor by participants of one faculty in a US university [F (2, 115) = 9.06, $p = 0.00$]. An LSD test showed that administrative staff evaluated this sub factor significantly higher

Table 4.32 Mean, SD, and F value of evaluation of administrative support

Country	Students		Lecturers		Staff		*F*	*P*
	M	*SD*	*M*	*SD*	*M*	*SD*		
AUS participants	3.46	0.50	3.30	0.47	4.00	0.53	5.68	0.005**
USA participants	3.55	0.57	3.39	0.47	4.15	0.37	9.06	0.00***

** $p < 0.01$; *** $p < 0.001$

than students and lecturers. However, students and lecturers evaluated this sub factor the same. The American students and lecturers believed administrative support was above average. On the other hand, American staff assessed administrative support as excellent.

Comparing the answers of participants of one faculty in an Australian university and one faculty in a US university showed that there were no significant differences in evaluation of the administrative support sub factor between Australian and American administrative staff [F (1, 20) = 0.60, p = 0.44]. Moreover, the results of ANOVA revealed that there was no significant difference in evaluation of this sub factor between Australian and American lecturers [F (1, 44) = 0.43, p = 0.51]. To continue, ANOVA test showed that there was no significant difference in evaluation of the sub factor between Australian and American students [F (1, 148) = 0.95, p = 0.33]. In both countries students and lecturers believed administrative support was above average. However, Australian and American staff believed administrative support was excellent.

4.3.5.2 Technical Support

Table 4.33 reports the means and standard deviations of the technical support sub factor based on the academic participants of one faculty in an Australian university and one faculty in a US university. As can be seen in this table, in Australia, the highest mean of the technical support sub factor belonged to lecturers (M = 3.50, SD = 0.94). After them, the administrative staff reported the sub factor (M = 3.37, SD = 0.51) as high and the lowest score was reported by students (M = 2.64, SD = 0.56). To investigate if there are any differences in evaluation of the technical support sub factor between students, lecturers and administrative staff, ANOVA was applied. The results showed that there was significant main effect of academic position on evaluation of it by participants of one faculty in an Australian university [F (2, 98) = 15.74, p = 0.00]. An LSD test revealed that students evaluated this sub factor significantly lower than lecturers and administrative staff. But administrative staff and students evaluated this sub factor the same. The results showed that Australian lecturers and staff believed technical support to be above average. However, Australian students believed technical support was only at an average level.

In one faculty in a US university, as can be seen in this table, the highest mean of answers to the technical support sub factor belonged to lecturers (M = 4.13,

Table 4.33 Mean, SD, and F value of evaluation of technical support

Country	Students		Lecturers		Staff		*F*	*P*
	M	*SD*	*M*	*SD*	*M*	*SD*		
AUS participants	2.64	0.56	3.50	0.94	3.37	0.51	15.74	0.00***
USA participants	3.15	0.62	4.13	0.72	4.07	0.64	27.92	0.00***

*** $p < 0.001$

SD = 0.72). After them, administrative staff reported this factor next highest (*M* = 4.07, *SD* = 0.64) and the lowest score was reported by the students (*M* = 3.15, *SD* = 0.62). To investigate if there are any differences in evaluation of the technical support sub factor between American students, lecturers and administrative staff, ANOVA was applied. The results showed that there was significant main effect of academic position on evaluation of this sub factor by participants of one faculty in a US university [F (2, 115) = 27.92, p = 0.00]. An LSD test showed that students evaluated the sub factor significantly lower than lecturers and administrative staff. There were no differences between evaluation of lecturers and administrative staff of the technical support sub factor. The results showed that American staff and lecturers believed research technical support was at an excellent level while the American students assessed technical support as above average.

Comparing the answers of participants of one faculty in an Australian university and one faculty in a US university showed that there were significant differences in evaluation of the technical support sub factor between Australian and American administrative staff [F (1, 20) = 6.82, p = 0.01] in that the Americans significantly evaluated this sub factor higher than Australians. Furthermore, the results of ANOVA revealed that there was significant difference in evaluation of this sub factor between Australian and American lecturers [F (1, 44) = 6.49, p = 0.01] in that the Americans evaluated this sub factor significantly higher than Australians. To continue, ANOVA test showed that there was also significant difference in evaluation of the technical support sub factor between Australian and American students [F (1, 148) = 26.87, p = 0.00] illustrating that American students evaluated the sub factor significantly higher than Australian students. Looking at the results, Australian staff and lecturers as well as American students had the same assessment namely that technical support practice was above average. However, Australian students believed technical support practice was only average. On the other hand, American staff and lecturers believed that technical support practice was at an excellent level.

4.3.5.3 Academic Support

Table 4.34 reports the means and standard deviations of the academic support sub factor based on the academic participants of one faculty in an Australian university and one faculty in a US university. As can be seen in this table, in Australia the highest mean of the academic support sub factor belonged to administrative staff

Table 4.34 Mean, SD, and F value of evaluation of academic support

Country	Students		Lecturers		Staff		*F*	*P*
	M	*SD*	*M*	*SD*	*M*	*SD*		
AUS participants	3.09	0.58	3.00	0.72	4.00	0.53	8.48	0.00***
USA participants	3.25	0.54	3.44	0.56	4.30	0.48	20.93	0.00***

*** $p < 0.001$

($M = 4.00$, $SD = 0.53$). After them, the students reported this sub factor ($M = 3.09$, $SD = 0.58$) as high and the lowest score was reported by lecturers ($M = 3.00$, $SD = 0.72$). To investigate if there are any differences in evaluation of the academic support sub factor between students, lecturers and administrative staff, ANOVA was applied. The results showed that there was significant main effect of academic position on evaluation of this sub factor by participants of one faculty in an Australian university [F (2, 98) = 8.48, $p = 0.00$]. An LSD test showed that administrative staff evaluated this sub factor significantly higher than students and lecturers. However, lecturers and students evaluated this sub factor the same. The results showed that the Australian students and lecturers believed academic support to be above average. However, Australian staff believed academic support was excellent.

In one faculty in a US university, as can be seen in this table, the highest mean of responses to the academic support sub factor belonged to administrative staff ($M = 4.30$, $SD = 0.48$). After them, lecturers reported this factor next highest ($M = 3.44$, $SD = 0.56$) and the lowest score was reported by the students ($M = 3.25$, $SD = 0.54$). To investigate if there are any differences in evaluation of this sub factor between American students, lecturers and administrative staff, ANOVA was applied. The results showed that there was significant main effect of academic position on evaluation of the academic support sub factor by participants of one faculty in a US university [F (2, 115) = 20.93, $p = 0.00$]. An LSD test showed that administrative staff evaluated the academic support sub factor significantly higher than lecturers and students. There were no differences between evaluation of lecturers and students. The American students and lecturers believed academic support was above average. On the other hand the American staff assessed academic support to be at an excellent level.

Comparing the answers of participants of one faculty in an Australian university and one faculty in a US university showed that there were no significant differences in evaluation of the academic support sub factor between Australian and American administrative staff [F (1, 20) = 1.86, $p = 0.18$]. However, the results of ANOVA revealed that there was significant difference in evaluation of this sub factor between Australian and American lecturers [F (1, 44) = 5.23, $p = 0.02$]; Americans evaluated this sub factor significantly higher than Australians. To continue, the ANOVA test showed that there was no significant difference in evaluation of this sub factor between Australian and American students [F (1, 148) = 2.88, $p = 0.09$]. Comparing the results, the participants of one faculty in an Australian and one faculty in a US university had same assessments based on their positions that is, in both countries, students and lecturers believed academic support was above average. However, Australian and American staff believed academic support was excellent.

4.3.5.4 Support e-Practice

Table 4.35 reports the means and standard deviations of the support e-practice factor based on the academic participants of one faculty in an Australian university and one faculty in a US university. As can be seen in this table, in Australia the highest mean of the support e-practice factor belonged to administrative staff ($M = 11.37$, $SD = 0.91$). After them, the lecturers reported the support e-practice factor ($M = 9.80$, $SD = 1.47$) as high and the lowest score was reported by students ($M = 9.21$, $SD = 1.19$). To investigate if there are any differences in evaluation of this factor between students, lecturers and administrative staff, ANOVA was applied. The results showed that there was significant main effect of academic position on evaluation of the support e-practice factor by participants of one faculty in an Australian university [F (2, 98) = 11.71, $p = 0.00$]. An LSD test showed that administrative staff evaluated this factor significantly higher than students and lecturers. However, students and lecturers evaluated this factor the same. The results showed that all participants of one faculty in an Australian university gave the same assessment, namely above average.

In one faculty in a US university, as can be seen in this table, the highest mean of responses to the support e-practice factor belonged to administrative staff ($M = 12.53$, $SD = 0.77$). After them, lecturers reported this factor next highest ($M = 10.96$, $SD = 1.09$) and the lowest score was reported by the students ($M = 9.96$, $SD = 1.09$). To investigate if there are any differences in evaluation of this factor between American students, lecturers and administrative staff, ANOVA was applied. The results showed that there was significant main effect of academic position on evaluation of the support e-practice factor by participants of one faculty in a US university [F (2, 115) = 35. 82, $p = 0.00$]. LSD test showed that administrative staff evaluated it significantly higher than lecturers and students. Also, lecturers evaluated this factor significantly higher than students. The results showed that American students and lecturers believed support practice was above average. However, the staff believed support practice was excellent.

Comparing the answers of participants of one faculty in an Australian university and one faculty in a US university showed that there were significant differences in evaluation of the support e-practice factor between Australian and American administrative staff [F (1, 20) = 9.71, $p = 0.006$]; Americans significantly evaluated it higher than Australians. Moreover, the results of ANOVA revealed that there was significant difference in evaluation of this factor between Australian and American lecturers [F (1, 44) = 9.29, $p = 0.004$]; Americans significantly

Table 4.35 Mean, SD, and F value of evaluation of support e-practice

Country	Students		Lecturers		Staff		*F*	*P*
	M	*SD*	*M*	*SD*	*M*	*SD*		
AUS participants	9.21	1.19	9.80	1.47	11.37	0.91	11.71	0.00***
USA participants	9.96	1.09	10.96	1.09	12.53	0.77	35.82	0.00***

*** $p < 0.001$

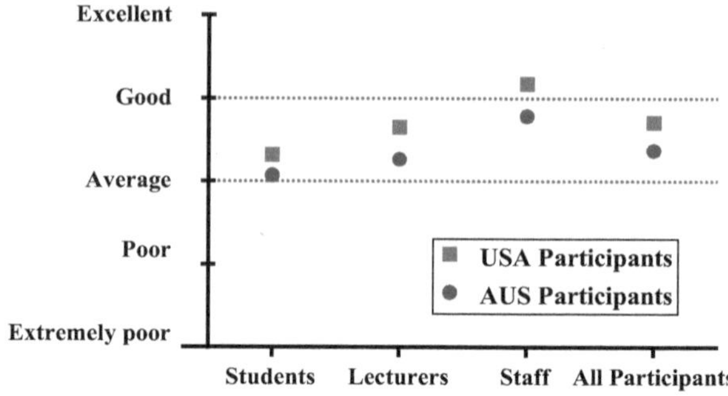

Fig. 4.5 Mean level of support e-practice

evaluated it higher than Australians. To continue, ANOVA test showed that there was significant difference in evaluation of this factor between Australian and American students [F (1, 148) = 16.01, p = 0.00] illustrating that American students evaluated it significantly higher than Australian students. Overall, comparing the results showed that all participants of one faculty in an Australian university and American students and lecturers had same perspective, believing that support practice was above average. However, the American staff believed support practice was excellent (see Fig. 4.5).

4.3.6 *Performance Appraisal Factor Results*

The performance appraisal factor was measured by 3 sub factors namely: cost effectiveness, learning effectiveness and satisfaction. In this section, the results of each sub factor based on participants' assessment are reported first, then the total results of all sub factors as the main factors of performance appraisal e-practice are reported.

4.3.6.1 Cost Effectiveness

Table 4.36 reports the means and standard deviations of the cost effectiveness sub factor based on the academic participants of one faculty in an Australian university and one faculty in a US university. As can be seen in this table, in Australia, the highest mean of the cost effectiveness sub factor belonged to administrative staff

Table 4.36 Mean, SD, and F value of evaluation of cost effectiveness

Country	Students		Lecturers		Staff		*F*	*P*
	M	*SD*	*M*	*SD*	*M*	*SD*		
AUS participants	6.14	0.94	5.90	1.11	6.37	0.91	0.78	0.45
USA participants	6.07	0.90	5.45	0.75	4.30	0.75	25.27	0.00***

*** $p < 0.001$

($M = 6.37$, $SD = 0.91$). After them, the students reported the sub factor ($M = 6.14$, $SD = 0.94$) as high and the lowest score was reported by lecturers ($M = 5.90$, $SD = 1.11$). To investigate if there are any differences in evaluation of this sub factor between students, lecturers and administrative staff, ANOVA was applied. The results showed that there was no significant main effect of academic position on evaluation of the cost effectiveness sub factor on participants of one faculty in an Australian university [F (2, 98) = 0.78, $p = 0.45$]. The results showed that Australian students and staff gave the same assessment, that is, they believed cost effectiveness was above average, while the Australian lecturers believed cost effectiveness was at an average level.

In one faculty in a US university, as can be seen in this table, the highest mean of answers to the cost effectiveness sub factor belonged to students ($M = 6.07$, $SD = 0.90$). After them, lecturers reported this factor next highest ($M = 5.45$, $SD = 0.75$) and the lowest score was reported by the administrative staff (M =). To investigate if there are any differences in evaluation of the cost effectiveness sub factor between American students, lecturers and administrative staff, ANOVA was applied. The results showed that there was significant main effect of academic position on evaluation of this sub factor by participants of one faculty in a US university [F (2, 115) = 25.27, $p = 0.00$]. LSD test showed that students evaluated the sub factor significantly higher than lecturers and administrative staff. Lecturers also evaluated it significantly higher than administrative staff. The American result showed that the staff and lecturers believed cost effectiveness to be average. On the other hand, the American students assessed cost effectiveness as above average.

Comparing the answers of participants of one faculty in an Australian university and one faculty in a US university showed that there were significant differences in evaluation of the cost effectiveness sub factor between Australian and American administrative staff [F (1, 20) = 31.80, $p = 0.00$]; Australians significantly evaluated this sub factor higher than Americans. However, the results of ANOVA revealed that there was no significant difference in evaluation between Australian and American lecturers [F (1, 44) = 2.55, $p = 0.11$]. To continue, ANOVA test showed that there was no significant difference in evaluation of the sub factor between Australian and American students [F (1, 148) = 0.17, $p = 0.67$]. Comparing the results, the Australian students and staff and the American students gave the same assessment that cost effectiveness was above average. However, Australian lecturers with American lecturers and staff believed cost effectiveness was only at an average level.

4.3.6.2 Learning Effectiveness

Table 4.37 reports the means and standard deviations of the learning effectiveness sub factor based on the academic participants of one faculty in an Australian university and one faculty in a US university. As can be seen in this table, the highest mean of the learning effectiveness sub factor in Australia belonged to administrative staff ($M = 12.37$, $SD = 1.59$). After them, the students reported this

Table 4.37 Mean, SD, and F value of evaluation of learning effectiveness

Country	Students		Lecturers		Staff		*F*	*P*
	M	*SD*	*M*	*SD*	*M*	*SD*		
AUS participants	11.23	1.42	10.65	1.08	12.37	1.59	4.50	0.01*
USA participants	11.0	1.19	10.9	0.91	12.3	0.94	8.16	0.00***

* $p < 0.05$; *** $p < 0.05$

sub factor ($M = 11.23$, $SD = 1.42$) as high and the lowest score was reported by lecturers ($M = 10.65$, $SD = 1.08$). To investigate if there are any differences in evaluation of the learning effectiveness sub factor between students, lecturers and administrative staff, ANOVA was applied. The results showed that there was no significant main effect of academic position on evaluation of this sub factor by participants of one faculty in an Australian university [F (2, 98) = 4.50, $p = 0.01$]. LSD test showed that administrative staff evaluated this sub factor significantly higher than students and lecturers, however, students and lecturers evaluated this sub factor the same. The results showed that the Australian students and lecturers had same assessment namely that they believed learning effectiveness was above average. However, the Australian staff believed learning effectiveness was excellent.

In one faculty in a US university, as can be seen in this table, the highest mean of the answers to the learning effectiveness sub factor belonged to administrative staff ($M = 12.30$, $SD = 0.94$). After them, students reported this factor next highest ($M = 11.00$, $SD = 1.19$) and the lowest score was reported by the lecturers ($M = 10.93$, $SD = 0.91$). To investigate if there are any differences in evaluation of the earning effectiveness sub factor between American students, lecturers and administrative staff, ANOVA was applied. The results showed that there was significant main effect of academic position on evaluation of the learning effectiveness sub factor by participants of one faculty in a US university [F (2, 115) = 8.16, $p = 0.00$]. LSD test showed that administrative staff evaluated the sub factor significantly higher than lecturers and students. There were no differences between the evaluation of lecturers and students. The same as the Australian result, the American students and lecturers believed learning effectiveness was above average. On the other hand, the American staff assessed learning effectiveness as excellent.

Comparing the answers of participants of one faculty in an Australian university and one faculty in a US university showed that there were no significant differences in evaluation of the learning effectiveness sub factor between Australian and American administrative staff [F (1, 20) = 0.01, $p = 0.90$]. Also, the results of ANOVA revealed that there was no significant difference in evaluation of it between Australian and American lecturers [F (1, 44) = 0.89, $p = 0.34$]. To continue, ANOVA test showed that there was also no significant difference in evaluation of this sub factor between Australian and American students [F (1, 148) = 1.23, $p = 0.26$]. Comparing the results, the participants of one faculty in an Australian and one faculty in a US university had the same assessment based on their position, that is, in both countries, students and lecturers believed learning effectiveness was above average. However, Australian and American staff believed learning effectiveness was excellent.

4.3.6.3 Satisfaction

Table 4.38 reports the means and standard deviations of the satisfaction sub factor based on the academic participants of one faculty in an Australian university and one faculty in a US university. As can be seen in this table, in Australia the highest mean of the satisfaction sub factor belonged to administrative staff ($M = 11.50$, $SD = 0.75$). After them, the students reported this sub factor ($M = 10.09$, $SD = 1.35$) as high and the lowest score was reported by lecturers ($M = 9.35$, $SD = 1.78$). To investigate if there are any differences in evaluation of the satisfaction sub factor between students, lecturers and administrative staff, ANOVA was applied. The results showed that there was significant main effect of academic position on evaluation of the satisfaction sub factor participants of one faculty in an Australian university [F (2, 98) = 6.66, $p = 0.002$]. LSD test showed that administrative staff evaluated this sub factor significantly higher than students and lecturers. Also students evaluated this sub factor significantly higher than lecturers. The results showed that all participants of one faculty in an Australian university gave the same assessment believing satisfaction to be above average.

In one faculty in a US university, as can be seen in this table, the highest mean of answers to the satisfaction sub factor belonged to administrative staff ($M = 12.46$, $SD = 0.96$). After them, lecturers reported this factor next highest ($M = 11.01$, $SD = 0.95$) and the lowest score was reported by the students ($M = 10.28$, $SD = 1.40$). To investigate if there are any differences in evaluation of the satisfaction sub factor between American students, lecturers and administrative staff, ANOVA was applied. The results showed that there was significant main effect of academic position on evaluation of satisfaction sub factor by participants of one faculty in a US university [F (2, 115) = 17.11, $p = 0.00$]. LSD test showed that administrative staff evaluated this sub factor significantly higher than lecturers and students. Also, lecturers evaluated it significantly higher than students. The results showed that the American students and lecturers believed that satisfaction was above average. However, the staff believed it was excellent.

Comparing the answers of participants of one faculty in an Australian university and one faculty in a US university showed that there were significant differences in evaluation of the satisfaction sub factor between Australian and American administrative staff [F (1, 20) = 5.71, $p = 0.02$] in that Americans significantly evaluated

Table 4.38 Mean, SD, and F value of evaluation of satisfaction

Country	Students		Lecturers		Staff		*F*	*P*
	M	*SD*	*M*	*SD*	*M*	*SD*		
AUS participants	10.09	1.35	9.35	1.78	11.50	0.75	6.66	0.002**
USA participants	10.28	1.40	11.01	0.95	12.46	0.96	17.11	0.00***

** $p < 0.01$; *** $p < 0.001$

the satisfaction sub factor higher than Australians. Moreover, the results of ANOVA revealed that there was significant difference in evaluation of this sub factor between Australian and American lecturers [F (1, 44) = 15.98, p = 0.00] in that Americans significantly evaluated the sub factor higher than Australians. However, an ANOVA test showed that there was no significant difference in evaluation of the satisfaction sub factor between Australian and American students [F (1, 148) = 0.69, p = 0.40]. Overall, comparing the results, all participants of one faculty in an Australian university and American students as well as lecturers had the same perspective, that is, they believed satisfaction was above average. However, the American staff believed satisfaction was excellent.

4.3.6.4 Performance Appraisal e-Practice

Table 4.39 reports the means and standard deviations of the performance appraisal e-practice factor based on the academic participants of one faculty in an Australian university and one faculty in a US university. As can be seen in this table, in Australia the highest mean of the performance appraisal e-practice factor belonged to administrative staff (M = 30.25, SD = 2.49). After them, the students reported the factor (M = 27.47, SD = 2.84) as high and the lowest score was reported by lecturers (M = 25.90, SD = 3.00). To investigate if there are any differences in evaluation of the performance appraisal e-practice factor between students, lecturers and administrative staff, ANOVA was applied. The results showed that there was no significant main effect of academic position on evaluation of the performance appraisal e-practice factor by participants of one faculty in an Australian university [F (2, 98) = 6. 76, p = 0.002]. LSD test showed that administrative staff evaluated this factor significantly higher than students and lecturers. Also, students evaluated this factor significantly higher than lecturers.

In one faculty in a US university, as can be seen in this table, the highest mean of responses to the performance appraisal e-practice factor belonged to administrative staff (M = 29.07, SD = 1.18). After them, lecturers reported this factor next highest (M = 27.39, SD = 1.57) and the lowest score was reported by the students (M = 27.36, SD = 2.27). To investigate if there are any differences in evaluation of this factor between American students, lecturers and administrative staff, ANOVA was applied. The results showed that there was significant main effect of academic position on evaluation of the performance appraisal e-practice factor by participants of one faculty in a US university [F (2, 115) = 3.98, p = 0.02]. LSD test showed

Table 4.39 Mean, SD, and F value of evaluation of performance appraisal

Country	Students		Lecturers		Staff		F	P
	M	SD	M	SD	M	SD		
AUS participants	27.47	2.84	25.9	3.00	30.25	2.49	6.76	0.002**
USA participants	27.36	2.27	27.39	1.57	29.07	1.18	3.98	0.02*

* $p < 0.05$; ** $p < 0.01$

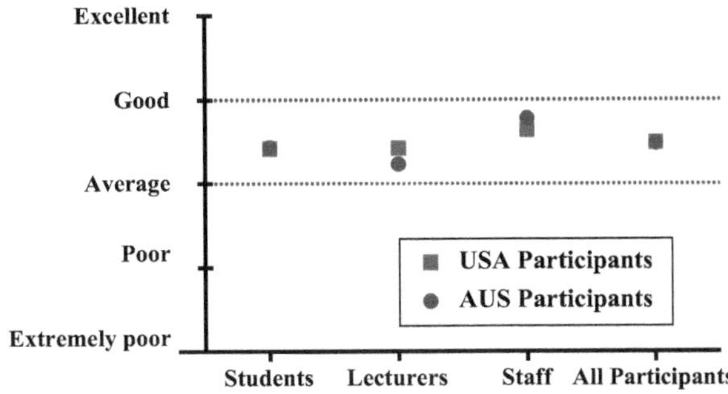

Fig. 4.6 Mean level of performance appraisal e-practice

that administrative staff evaluated the e-practice factor significantly higher than lecturers and students. There were no differences in evaluation of lecturers and students of the factor. Comparing the answers of participants of one faculty in an Australian university and one faculty in a US university showed that there were no significant differences in evaluation of the performance appraisal e-practice factor between Australian and American administrative staff [F (1, 20) = 2.14, $p = 0.16$].

However, the results of ANOVA revealed that there was significant difference in evaluation of this factor between Australian and American lecturers [F (1, 44) = 4.62, $p = 0.03$]; Americans evaluated this factor significantly higher than Australians. To continue, ANOVA test showed that there was no significant difference in evaluation of the factor between Australian and American students [F (1, 148) = 0.07, $p = 0.78$].

Overall, it seems that all participants of one faculty in an Australian and one faculty in a US university had the same assessment, believing performance appraisal was above average (see Fig. 4.6).

4.3.7 E-Practice Results

As can be seen in Table 4.40, in Australia the highest mean of the total e-practice factor belonged to administrative staff ($M = 163.37$, $SD = 6.06$). After them, lecturers ($M = 145.30$, $SD = 13.75$) reported the total as high and the lowest score was reported by students ($M = 144.85$, $SD = 11.30$).To investigate if there are any differences in evaluation of the total e-practice factor between Australian students, lecturers and administrative staff, ANOVA was applied. The results showed that there was significant main effect of academic position on evaluation of the total

Table 4.40 Mean, SD, and F value of evaluation of e-practice

Country	Students		Lecturers		Staff		*F*	*P*
	M	*SD*	*M*	*SD*	*M*	*SD*		
AUS participants	144.85	11.30	145.3	3.00	163.37	13.75	9.37	0.00***
USA participants	153.54	7.57	156.5	5.60	158.07	5.54	3.41	0.03*

* $p < 0.05$; *** $p < 0.001$

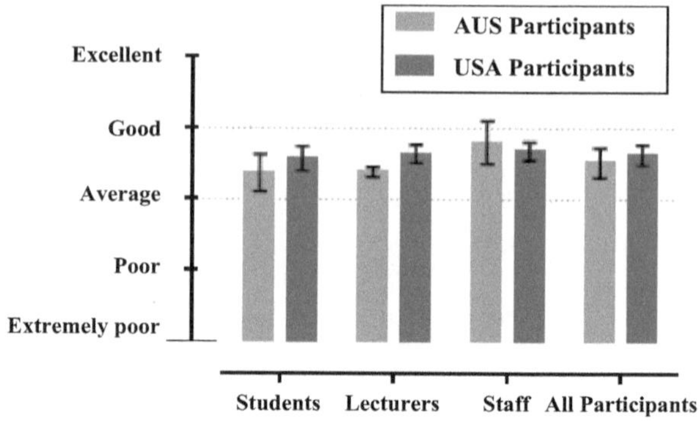

Fig. 4.7 Mean (and SD) level of e-practice

e-practice factor participants of one faculty in an Australian university [F (2, 98) = 9.37, p = 0.00]. LSD test showed that administrative staff evaluated the total e-practice factor significantly higher than lecturers and students.

In America the highest mean of the total e-practice factor belonged to administrative staff (M = 158.07, SD = 5.54). After them, the lecturers reported this factor (M = 156.51, SD = 5.60) as high and the lowest score was reported by students (M = 153.54, SD = 7.57). To investigate if there are any differences in evaluation of the total e-practice factor between American students, lecturers and administrative staff, ANOVA was applied. The results showed that there was significant main effect of academic position on evaluation of the total e-practice factor by participants of one faculty in a US university [F (2, 115) = 3.41, p = 0.03]. LSD test showed that administrative staff evaluated the total e-practice factor significantly higher than lecturers and students.

Finally, the total scores of e-practice in both countries were analysed. There were no significant differences on evaluation of the total e-practice factor between Australian and American administrative staff [F (1, 20) = 4.21, p = 0.05]. However, the results of ANOVA revealed that there was significant difference in its evaluation between Australian and American lecturers [F (1, 44) = 13.83, p = 0.001]; Americans evaluated this factor significantly higher than Australians. Further, an ANOVA test showed that there was significant difference in evaluation of the total e-practice factor between Australian and American students [F (1, 148) = 30.85, p = 0.00]; American students evaluated it higher than Australians. In general, both countries evaluated total e-practice above average (see Fig. 4.7).

This chapter has investigated the current status of e-practice in Australian and American universities. The results showed that the level of e-practice in all factors were above average in both countries. However, participants of one faculty in a US university assessed e-practice elements higher than Australian participants. The following chapter aims to investigate the current issues and problems of e-learning practice in 4 aspects, namely pedagogy, culture, technology and e-practice, to identify which needs to improve. It does so by applying a qualitative method.

Chapter 5
The Current Issues Concerning E-Practices

5.1 Introduction

The third aim of this research program was to investigate current issues in the e-learning practices of in one faculty in an Australian university and one faculty in a US University. To do so, the researcher conducted a study into four aspects of e-learning, namely: (1) pedagogy, (2) culture (3) technology, and (4) the aspects of e-learning status in order to identify areas which most need to be improved upon.

As has been explained earlier in chapter three on methodology, interviews with students, lecturers and administrative staff involved in the e-learning educational system were conducted. These data were coded, categorized and analyzed by the researcher using Nvivo in data analysis. Each area mentioned above was covered in the research questions asked in the interviews. Within each of these areas, questions were asked that focused on different facets of the area.

Four main questions were asked about pedagogical issues: the approach towards e-learning that the participants apply in their online educational system; the effective learning practice that participants have chosen that reflects their approach to e-learning; the e-learning assessment methods that participants have engaged with; and the tools and instruments that are supplied for the learning of the educational content within an e-learning environment. Based on participants' responses to these questions, thirteen themes have been coded from the Australian responses and fourteen themes from the American responses.

For the investigation of cultural status and issues, the researcher asked three main types of questions. Questions that highlighted the existence of cultural issues and identified sensitive areas; questions about the non-existence of cultural issues; and questions about effective cultural practices. As will be explained in the results section, the researcher coded eight themes from participants of one faculty in an Australian university and the same eight themes from participants of one faculty in an American University.

S.H. Sadeghi, *E-Learning Practice in Higher Education: A Mixed-Method Comparative Analysis*, Studies in Systems, Decision and Control 122,
https://doi.org/10.1007/978-3-319-65939-8_5

To investigate the current status of technological issues and challenges, the researcher asked three main questions. These concerned the current status of the technology, the level of the infrastructure of technology and the level of the function of the e-learning educational system. It is important to note that each response was categorised in three levels of importance—low, moderate and high. The responses of participants of one faculty in an Australian university have been coded with six themes; however, the responses of participants of one faculty in a US University have been coded with seven themes (See results section for more detail).

To investigate the best aspects of e-learning educational systems and the areas to improve in both countries, the researcher asked two main questions: about the best aspects of e-learning and about the challenges and problems that need to be addressed. According to the answers of interviewees, the researcher coded nine themes from Australia and coded eight themes from the American participants.

To summarize, the investigation of the four main research areas utilized eleven interview questions to both the Australian and the American participants. Thirty-seven themes were coded from the Australian data and thirty-six from the American data.

5.2 Methodology

The methods selected for this research were a qualitative approach and a case study design. Broadly, the qualitative approach was considered to be well suited to the aim of the research. Findings obtained from qualitative approaches such as interviews enable a researcher to present a picture of reality and reveal its complexities. Qualitative data provide open-ended information (Creswell and Plano Clark 2007) from which inductive logic can be used to reveal an emerging pattern (Creswell 2013) in contrast to statistical or quantitative approaches which only provide superficial understandings of the issue.

The research strategy of case study was adopted because the study sought to explore the current status of e-learning in two specific contexts in two different countries. Case studies can provide data that are thick, rich and descriptive (Ellinger et al. 2005). Maximum variation sampling (Patton 2002) was used in selecting the two cases. This sampling strategy is useful to "document unique or diverse variations that have emerged in adapting to different conditions and contexts" and to "identify important common patterns that cut across variations" (Patton, p. 243).

5.2.1 Participants

Purposeful sampling was used to select the participants. Purposeful sampling is one in which the researcher deliberately selects participants "that differ on some characteristics or trait" (Creswell 2013, pp. 207–208) The 29 participants were students, lecturers and administrative staff of two universities, one in Australia and one in United States. The participants were selected because they were assumed to have familiarity with the e-practices of their universities. They were chosen based on their availability and willingness to participate. To assist the comparability of the two groups, the participants were drawn from schools/faculties that were roughly equivalent, namely in the area of health sciences/public health (Table 5.1).

5.2.2 *Data Collection Approach*

The framework of the research was based on Maxwell's data-planning matrix. See Table 5.2 for details. The study used semi-structured interviews as the data collection process. Semi-structured interviews are considered a suitable approach to investigate the issue under consideration in qualitative research.

According to Denscombe (2014), semi structured interviews allow "the interviewee to develop ideas and speak more widely on the issues raised by the researcher" (p. 176). The type of questions asked is also critical. Patton (2002) suggested that questions, to be comprehensive, should cover the experience, opinions, feeling, knowledge, sensory effects and demographic details of the interviewees. Consequently this study, which aimed to compare the current status of e-practice in two countries, designed the questions along those lines. Table 5.3 outlines the type of questions used by the researcher.

The data were collected from students, lecturers and administrators at one Australian university and one American University. A pilot study was carried out in February, 2014. Feedback from this allowed the researcher to identify potential misunderstandings. The wording and structure of the research instruments was then refined. The interviews with the participants of one faculty in a US University were conducted via Skype and phone between February and June 2014. Interviews with participants of one faculty in an Australian university were conducted in an office on campus, or in the participants' homes, or via Skype or phone. The interviews generally lasted around 20 min.

Table 5.1 The number of participants in the pilot study and data collection

Participants of interview	USA participants			AUS participants			Total number
	Pilot study	Data collection	N	Pilot study	Data collection	N	
Faculty	0	3	3	1	8	9	12
Admin	0	1	1	0	2	2	3
Student	2	7	9	1	4	5	14
Total number	2	11	13	2	14	16	29

Table 5.2 Data-planning matrix

What do I need to know?	Why do I need to know this?	What kind of data is obtained question?	Where can I find the data?	Timeline 2014
Current status of e-practice focused on:	To identify, describe and compare the pedagogical, cultural and technological requires, challenges and issues concerning e learning practices	Semi-structured interviews	An Australian university	Feb to June
1-Pedagogical issues			An American University	
2-Cultural issues and sensitivity				
3-Technological challenges				
4-Best aspects of e-practices				
5-Areas which need to improvement				

Table 5.3 The current statues of e-practices and type of questions

Type of questions	Pedagogical issues	Cultural issues	Technological issues	Best and improve
Experience	1 Q	1 Q	1 Q	1 Q
Opinion	2 Q	1 Q	2 Q	1 Q
Demographic	6 Q			

5.2.3 *Data Analysis*

Data gathered from the interviews were recorded, transcribed and then coded using NVivo. The process of data analysis involves "organization, classification, categorization, a search for patterns, and synthesis" (Schloss and Smith 1999, p. 190). In order to facilitate the emergence of key themes, and due to the comparative nature of the study, the constant comparative method (CCM) developed by Glaser and Strauss (Glaser and Strauss 1967; Strauss 1987; Glaser 1992), was the system of analysis selected. The process involved 6 stages.

Stage 1 Data collection. In this study, information concerning pedagogy, culture, technology and the best aspects of e-learning in the 2 universities was obtained from the participants via interview.

Stage 2 Identify important issues; use them to create categories. This stage of data analysis involved the use of NVivo software for coding the relevant data from different sources. Broad categories (pedagogical practice required, pedagogical challenges, best aspects of e-practices and e-practice areas which need improvement, organizational dilemmas) were pre-generated based on foci of the research questions and results of study 1 in Chap. 3 and study 2 in Chap. 4. This is a process called "a priori codes" (Miles and Huberman (1994).

Stage 3 Collect additional data; elaborate on dimensions within categories. In this study, no additional data were needed as data already generated were adequate. Sub-categories were generated within the broader categories.

Stage 4 Describe the categories; reformulate and delete as necessary. Several of the initial broad categories (pedagogical practice required, cultural practice issues, the current levels of technological practice, key technological challenges, best aspects of e-practices and e-practices which need to be improved) were found to be supported, with the emergence of recurring instances from repetitive and careful reading of the data. Similar findings did not occur for two initial broad categories (pedagogical challenges and organizational dilemmas), so these categories were deleted.

Stage 5 Identify patterns and relationships. The idea is to "build up to patterns, theories and generalizations" (Creswell and Plano Clark 2007, p. 24). This process of interpretation was enacted by the researcher and is described in this chapter.

Stage 6 Develop a theory while comparing and refining data. In this study, an e-practice model of current status was developed, supported by e-practice assessment results and dominant aspects of cultural dimensions and learning preferences. In conclusion, triangulation was used throughout the entire data analysis process. Triangulation of data validates the accuracy of the findings (Creswell 2013) and increases their credibility (Schloss and Smith 1999).

5.3 Research Demographics

There was a total of thirteen interviewees (10 females = 77% and 3 males = 23%) who participated in the American research and a total of sixteen interviewees (eleven females = 68.75% and 5 males = 31.25%) who participated in the

Table 5.4 Demographic information based on gender

Country	Gender	Number	Mean	Percent
USA participants	Female	10	6.00	44.82
	Male	3		
AUS participants	Female	11	8.00	55.18
	Male	5		
Total		29		

Australian study. Table 5.4 is a descriptive summary of the gender makeup of the interviewees.

Table 5.5 shows the demographic information based on the age of the interviewees and the position they hold within the university. Age has been divided into four groups: 20–30, 30–40, 40–50 and 50–60 years. Although there were no interviewees in 40–50 year group in the American sample, 46.15% of participants belonged to the 20–30 year group. In contrast there were no interviewees in the 20–30 year group in the Australian sample; however, 43.75% of participants belonged to the 30–40 year age group. In regards to the position held within the university, most of the participants in the American sample were students (N = 9; 69.22%), however, in the Australian sample most of the participants were lecturers (N = 9; 56.25%). It is worth noting that in total 41.38% of participants were students, 48.27% were lecturers and 10.35% were administrative staff.

Table 5.6 shows the field of study of participants and their level of experience in an e-learning environment. In this study, research has focused on the health and medical science fields. The reason for this selection is that there are so many different online courses in those countries and there are different and specific online courses in each country that differ from each other. It was also found in the pilot study that most of the online and joint courses were in the health and medical sciences field. Indeed the researcher selected for this sample study from environmental health (n = 2), epidemiology (n = 5), health informatics (n = 1), biostatistics (n = 1) and health system (n = 4) in American sample. Also in Australia the selected sample is from occupational therapy (n = 2), speech pathology (n = 2), health system (n = 10) and health science research methods (n = 2).

The experience of the participants has been categorised into the following areas; teaching, designing and programming and online study. According to these

Table 5.5 Demographic information based on age and position

Participants	Age	N[a]	Percent of participants	Position	N[a]	Percent of participants
USA	20–30	6	56.15	Faculty	3	23.08
	30–40	4	30.77	Staff	1	7.70
	40–50	0	00.00	Student	9	69.22
	50–60	3	23.08			
	Total				13	44.83
AUS	20–30	0	00.00	Faculty	9	56.25
	30–40	7	43.75	Staff	2	12.5
	40–50	5	31.25	Student	5	31.25
	50–60	4	25.00			
	Total				16	55.17
Total					29	

Table 5.6 Demographic information based on online course experience

Participants	Online course area	N[a]	Experience	N	Level of experience	N	Total
USA	Environmental health	2	Teaching	2	Beginner (1–3 course)	3	
	Epidemiology	5	Designing	2	Average (4–6 course)	3	
	Health informatics	1	Online study	8	Expert (over 6 course)	7	
	Biostatistics	1					
	Health system[b]	4					
	Total						13
AUS	Occupational therapy	2	Teaching	8	Beginner (1–3 course)	6	
	Speech pathology	2	Designing	3	Average (4–6 course)	2	
	Health system	10	Online study	5	Expert (over 6 course)	8	
	Research method	2					
	Total						16
Total							29

[a] N = No. of participants
[b]Health system area including Sociology, Strategic Communication, Physiotherapy and Veterinary Health system

findings, 61.54% of participants of one faculty in an American University had experience in online studies while 44.83% of participants of one faculty in an Australian university had experience in that area. Of the total sample, 46% had experience in online studies, 36% had experience in online teaching and 18% had experience in online design and programming.

The level of their e-learning experiences has been divided into three categories; from beginners to those with average experience to experts. The parameters for these categories are based on the number of online courses that the participants have engaged in. Indeed, most of the participants (15 people) in both countries reported their level of e-learning experience as expert (=51.72%). 31.03% reported their level of e-learning experience as beginner with one to three courses and 17.24% reported their level of e-learning experiences as average with three to six courses (see Table 5.6 for more information).

5.4 Key Finding of the Current Issues Concerning E-Practices

5.4.1 Pedagogical Issues

As has been mentioned in the introduction, the first issue that the researcher focused on was pedagogical approach. Table 5.7 illustrates the content of each question, the coded answers of the interviewees to these questions, the frequency of the participants' comments and the percentage of their responses in Australia and United States.

According to Table 5.7, the most comments recorded are in relation to assessment methods in the Australian interviews (n = 31; 34.83%), which was also the case in the American interviews (n = 27; 36.48%). This means that the participants had the most concerns about assessment methods within their e-learning system in both countries. According to these results the researcher found that they currently have four main issues in assessment methods. These are: assignments, group thread discussions, case based projects and self-assessments.

Table 5.7 Participant's comments on pedagogical issues based on country

Interview questions focused on	Participants	Theme	No of percentage
The required approaches to learning	AUS	Collaborative learning	23 (25.84)
		Feedback-based practices	
		Problem based learning	
	USA	Collaborative learning	18 (24.32)
		Outcome oriented	
		Feedback-based practices	
		Problem based learning	
Effective learning practice required	AUS	Online tools and activities	16 (17.97)
		Team-work	
	USA	Assessment	14 (18.91)
The required methods to assessment	AUS	Assignments	31 (34.83)
		Group thread discussions	
		Case based-project	
	USA	Self-assessment	27 (36.48)
The required learning content	AUS	Multimedia	19 (21.34)
		Online module	
	USA	Text	15 (20.27)

The question with the next highest frequency of comments is about the approaches to learning in the Australian interviews (N = 23; 25.84%), which was also the case in the American interviews (n = 18; 24.32%). The responses of interviewees have been categorized into three main themes; collaborative learning, outcome oriented, feedback based practices and problem based learning.

Nineteen (21.34%) comments by participants of one faculty in an Australian university and fifteen (20.27%) comments by participants of one faculty in a US University related to the content of e-learning. The responses have been categorized into the following areas; multimedia, online module and text.

Sixteen comments (17.97%) in Australia and fourteen (18.91%) in the USA reported learning experience required by the system. Their answers to this question have been categorized as online tools and activities, team work and assessment.

5.4.2 *The Required Approaches to Learning*

5.4.2.1 Australia Case Study

Investigation into the current status and the requirements of e-learning practice approaches in Australia is concerned with the policies of the educational system and the perspectives of suppliers including administrative staff, lecturers and students, as to the process of learning in e-learning courses.

According to the results derived from the interview responses, there are concerns that are focused on three themes; collaborative learning (N = 9; 39.13%), feedback based practice (N = 8; 34.78%) and problem based practice (N = 6; 26.08%). These results from Table 5.8 show that the first priority of the participants, regardless of their position, is collaborative learning rather than individual learning. Further, their references to this theme indicate that the type of collaboration and the interaction between students and lecturers are the key elements of engagement with the process of learning.

One lecturer said, "Engaging students in activities and discussions will encourage them to explore and understand material before class and the students will be challenged during online class." (JuS.F.1) A health science e-learning instructor also commented on this: "I use wikis for the collaborative learning process and expect students to engage online in an efficient and supportive fashion. This requires me to carefully construct the design of the learning opportunities and to be responsive as it rolls out" (MiC.F.5).

As an online student of public health management explained: "Collaborative and interactive strategy is highlighted in my classes because it is very interesting. The students always will have a great time during these units and there will be positive feedback activities from each other and teachers to learn via collaborative and interactive ways" (LeA.S.14).

In conclusion, it is apparent that the pedagogical perspective towards strategies tends to favour a more collaborative approach. In this case, one of the senior

Table 5.8 The required approaches to learning in Australia

Theme	Comments of participants	Source[a]	Percent
Collaborative learning	Engaging students in activities and discussions	9	39.13
	Collaborative and interactive online teaching		
	Collaborative learning process		
	Flexible co-active approach		
	Collaborative discussion approach		
	Collaborative and interactive		
	Collaborative flexible strategies		
	Collaborative and interactive strategies		
	Engaging with classmates and course collaborative		
Feedback-based practices	Detailed feedback tasks	8	34.78
	Feedback-oriented practices		
	Assignment centred and information giving		
	Quiz and feedback on assignments		
	Feedback practices		
	Feedback-oriented		
	Practices feedback-oriented		
	Positive feedback activities		
Problem based learning	Problem solving techniques	6	26.08
	Clinical problem based		
	Problem-based		
	Online problem based learning		
	Learner-centred problem-solving		
	Problem-based		

[a]N = 23

lecturers of physiotherapy teaching more than six online courses explained that: "My online teaching strategy is based on a flexible co-active approach and I also try to engage students to be active participants in the online learning environment. With the discussion activities, it is my main request" (PaU.F.7).

The results show that the second priority of participants is a feedback based approach. This approach refers to the tasks, quizzes, practices and activities that have been designed for online courses and the probable questions arising from them which constitutes a quick and easy method of learning. Some comments include: "From what I have seen in the WIL (Work Integrated Learning) unit in the discipline of occupational therapy (OT) the online teaching strategy needs more feedback-oriented practices" (MeR.F.4). "Apply to one of my units, which is fully online. The strategy is needed to present students with tasks as close to "real assignments and fast feedback" as possible, provide materials and examples about how to do the tasks, quiz and feedback on assignments" (RoB.F.8).

The third priority of participants regarding the pedagogical aspects refers to problem-based learning. This approach has focussed on problem solving techniques and is clinical problem based. As can be seen in Table 5.8, six comments are related to this approach in online courses. For example one of the senior lecturers noted, "Online problem based learning is the best strategy for my teaching, in which I am able to do constructivist teaching with students" (RoD.F.9). Related to this, an administrator who supports students explained that: "I have some administrative experience with blended and fully online education with postgraduate coursework. within our protocol guidelines. We need to focus on learner-centred problem-solving flexible strategy" (JaN.A.10).

5.4.2.2 USA Case Study

Four themes emerged from the interviews with regards to the approach to e-learning in the USA. These were collaborative learning (n = 9; 50.00%), outcome oriented approach (n = 4; 22. %), problem-based learning (n = 3; 17.00%), and feedback based practice (n = 2; 11.00). Table 5.9 summarizes the references that interviewees made that lead to these themes being categorized.

Table 5.9 The required approaches to learning in USA

Theme	Reference	Source[a]	Percent
Collaborative learning	Interaction between students and discussion boards	9	50.00
	Interactive		
	Collaborative based		
	Learner-centred interactive		
	Collaborative and interactive		
	Collaborative-centred		
	Group work; leadership, collaboration, dialog		
	Collaborative and interactive with the team		
	Interactive web-based program		
Outcome oriented	Outcome-oriented	4	22.00
	Highly focused classes were on student outcomes		
	Focused on student achievement		
	Online quizzes for better outcome		
Problem based learning	Problem-based assignments	3	17.00
	Problem solving		
	Problem-based		
Feedback-based practices	Feedback on assignments	2	11.00
	Feedback-oriented		

[a]N = 18

The theme with the highest frequency of comments by participants when explaining the current e-learning practice approach was collaborative learning (n = 9; 50.00%).

In this case participants believed that collaboration and interaction between students and between students and lecturers in discussion boards which improve teamwork is the first priority of the e-learning practice approach. Comments highlighted classes as being "process-oriented but transitioning to having more interaction between students" (Nat.F.1); as well as being "interactive and learner-centred" (Roc.S.5) and "collaborative based and focused on understanding the concepts being taught" (Dra.S.9). There is acknowledgment that classes "require group work; leadership, collaboration, where dialogue is key", (Han.S.8) although some students still experience courses that are interactive and collaborative alongside courses that are primarily in a text format (Cyn.F.13).

The next most frequently mentioned e-learning practice approach reported in America was an outcome oriented approach (n = 4; 22.20%). From this perspective, the outcome of learning is of most concern in educational practice. Comments like" highly focused classes were on student outcomes", "focused on student achievement" and "online quizzes for better outcome" have been reported by interviewees. For example one student noted that he "believes they are flexible, learner-centred and that the learning strategies are focused on student achievement" (Hea.S.7). One of the instructional designers said: "It is highly focused on student outcomes. Outcomes are established by the institution" (UofM, SPH), with "accrediting bodies and professional organizations, who establish core competencies for public health professionals" (Joe.A.3).

The third e-learning practice approach that was mentioned by participants of one faculty in an American University was problem-based learning (n = 3; 17.00%). In this case participants reported that practices like assignments based on problem solving were their dominant e-learning practice approach. Representative comments include: "In my opinion problem solving and discussion is the best strategy for e-learning courses" (Hel.S.6), "we have to focus on problem-based assignments where regular homework was assigned for students to complete. These assignments were promptly graded with feedback on how to improve. Usually, these homework assignments directly related to the material covered on the exams" (Isa.S.11).

The fourth e-learning practice approach that was indicated by interviewees was feedback based practice (n = 2; 11.00%). As well as feedback from the lecturers some classes have required students to interact and provide feedback to one another. They explained: "the online module provided different ways to learn the concepts with audio recording, visual clips, readings, research paper for understanding of application with feedbacks from the instructor and TA's." (Yin.S.10) and "most of my online classes have provided feedback on assignments and I appreciate that. Most of the times we also need to include a response to another student's comment. In this way, online classes require much more interaction and feedback from the student than from an in-class environment where each student doesn't necessarily need to talk in class if they don't want to" (Han.S.8).

5.4.2.3 The Results of Comparative Investigation

According to the results in Australia and America there are some similarities and differences in the perspective of participants about their main required approaches to learning. Firstly, according to results as can be seen in Table 5.10, it has been revealed that the issue of collaborative learning is the same in both countries. Although the percent of Americans (50%) who believed in this issue is higher than Australians (39.1%), learning with others has been the first consideration of both cultures. The second pedagogical issue which was similar in both countries was problem based learning. In this case, participants in both countries explained that the process of problem solving in their e-learning system needs to be improved because of the challenge that they encounter with it, although it seems that this requirement is more highlighted by Australians (26.1%) rather than Americans (17%). The third similarity in participants' concerns about approaches to learning issues refers to feedback-based practices. According to this theme 34.8% of Australians and 11% of Americans claimed that giving feedback in learning via assignment to students is another challenge they have to figure out. This requirement of course is in relation to the process of assessment learning in students that has to be considered by e-learning course providers. Regardless of these similarities about the main issues of pedagogical perspective, one difference has been found namely outcome oriented. This theme was reported as the second most important issue by Americans; however, Australians didn't report any concern about the outcome of learning. This difference between the perspective of Americans and Australians about the outcomes of learning can indicate that learning in America tends to be as practical and applicable as can be, however, this issue has been neglected by Australian course providers.

Table 5.10 Comparative result based on the required approaches to learning

The required approaches to learning	Percentage of comments		Comparison based on	
	AUS P[a]	USA P	Require	Priority
Collaborative learning	39.1	50	Similar	Similar
Outcome oriented	0	22	Different	Different
Problem based learning	26.1	17	Similar	Different
Feedback-based practices	34.8	11	Similar	Different

[a]Participants

5.4.3 *Effective Learning Practice Required*

5.4.3.1 Australia Case Study

Investigation of effective learning practice enquired into what participants have been supplied with within the e-learning system, and resulted in the emergence of three key themes. Participants focused on online tools and activities, team work and assessment. Table 5.11 summarizes Australian results. As can be seen, the first priority of participants' responses was about online tools and activities that they have to engage with in an e-learning system (n = 7; 43%).

According to Australian findings, tools like digital didactic material and applications, online wiki tool, secretive applications and tools for cheating, multiple online apps for alternative sources of information, blackboard discussion board, online tutorial and LMS have been supplied as effective learning practice required in e-learning system. For example a clinical educator highlighted this: "I think using secretive applications and tools for cheating that are able to check for cheating and something else on-line and off-line." (MiM.F.6)

Also two of the online teaching senior lecturers focused on online apps and LMS and commented on this: "In my opinion, fully online learning activities are convenient for students and teachers in the education system. As detailed in the total

Table 5.11 Effective learning practice required in one faculty in an Australian university

Theme	Comments of participants	Source[a]	Percent
Online tools and activities	Digital didactic material and applications	7	43.75
	Online wiki tool		
	Secretive applications and tools for cheating		
	Multiple online app alternative sources of information		
	LMS		
	Blackboard discussion board		
	Online tutorial		
Team-work	Stimulate the reflective team based learning cycle	5	31.25
	Learning culture within the group		
	Discussions and group projects		
	the collaborative learning strategy		
	team work based strategy is fabulous experience		
Assessment	Linking assessment with the activity	4	25.00
	Assignments and quiz		
	Practice quizzes		
	Students self-assessment of assignments		

[a]N = 16

online section of the faculty of health sciences website, LMS offers a range of online teaching chances in the health sciences disciplines, which are flexible and interactive experiences for learners both locally and globally" (RoD.F.9). "I provide students with multiple online app alternative sources of information, and have them apply the information, and provide detailed feedback about their performance" (RoB.F.8).

One of the administrators of online courses focused on "blackboard discussion board and also strategic discussion with experts in the field of pedagogical facilitating and technological tools" (MeG.A.11). One of the health science postgraduates noted, "The replacement of the person tutorial for the online tutorial improves the ability to reflect on differing outlooks, properly retain and consider well informed opinions of others and the thought processes that lead them to their conclusions" (SuS.S.16).

The second priority that has been put forward as an effective learning practice required in e-learning system and online courses is team work (n = 5; 31.25%). In this case comments like learning cycle, learning groups; group project, collaborative learning strategy and team work based strategy have been introduced by postgraduates, administrative staff and lecturers. Some comments were: "I used online learning extensively at postgraduate level. My pedagogy was adult learner centred; using authentic learning experiences in order to stimulate the reflective team based learning cycle" (MeR.F.4). "Residential sessions at the beginning of an on-line UoS (Unit of study) to forge a learning culture within the group, and to set up identification with the degree program and student cohort" (JaN.A.10). "The collaborative and team learning strategy is an effective learning strategy in that it can be a strong strategy for student engagement in online units" (GuN.S.13).

The third priority of participants regarding effective practice required in an e-learning educational system was assessment (N = 4; 25%). In this case they mentioned linking assessment with the activity, assignment and quiz, practice quizzes and self-assessment and assignments as effective strategic experiences that they used. One of the associate professors stated, "We have many options that we should be thinking about. One of the main things to do in the online environment in my experience with three online Health Sociologist units is linking assessment with the activity and focusing on flexible timeline for quiz." (KaR.F.2). Other perspectives for effective practice that are required are "based on assignments and quiz that are practical and clinically relevant" (MaR.F.3) and also "practice quizzes in the online environment" (PaU.F.7). While one of the postgraduates focused on students' self-assessment of assignments and stated, "I think self-assessment is a very effective process in my postgraduate online Public Health course" (LeA.S.14).

Participants focused on online tools and activities, team work and assessment. Table 5.6 summarizes Australian results.

5.4.3.2 USA Case Study

According to the American results the second pedagogical issue was the need for effective learning experience. One of the lecturers noted; "an effective practice is dependent on strategic planning methods of teaching and learning process, so in my opinion concise instructions on what is expected, deadlines and quick responses to inquiries is needed." (Nat.F.1) In this case, three main effective learning practices were mentioned by participants namely: online tools and activities, teamwork and assessment. Some of the participants focused on several effective experiences and explained that: "Assessment methods used were learning lab activities and projects, papers and presentation, discussion board quizzes and self-assessment" (Yin.S.10), As well as "…homework-reports and essays, online threaded discussions, presentations and webinars." (Dra.S.9)

Table 5.12 has the results concerning effective learning practice themes that participants considered are required. According to this table the first effective practice that was reported by participants was supplying online tools and activities to teach and learn (n = 6; 42.85%). Comments indicate that participants have used the following tools and activities: discussion boards, readings alongside online lectures, videos and animations as well as other types of online activities. What some have indicated is that they like a variety of tools that support one another: "The readings and the online lectures touch upon the same topics, which is very helpful for me. I learn better by receiving the same material in different forms: i.e. verbal vs. written." (Hea.S.7) and they "enjoy online methods that use a variety of formats to

Table 5.12 Effective learning practice required in one faculty in a USA University

Theme	Reference	Source[a]	Percent
Online tools and activities	The online discussion boards	6	42.85
	The e-readings tools and the online lectures		
	Videos and animations		
	HINF 5430 (Health Informatics online practices)		
	The various multimedia tools		
	Audio and video		
Team-work	Deadlines and quick responses in the small group work	6	42.85
	Engage on student-instructor, student-student		
	Group assignments and students learning by student teams		
	Group projects		
	Team collaboration		
	The Island Project		
Assessment	Homework assignments	2	14.30
	Work for the assignments		

[a]N = 14

teach. For example, provide audio, video, some reading (not too heavy on the reading!) and different kinds of online exercises." (Han.S.8). Another noted that being able to call into the course live was an effective learning practice (Roc.S.5).

The second of the effective practices that has been indicated by students, faculty and administrators of health science at the American University was team work (n = 6; 42.85%). According to the responses, interacting with others within the educational system to find quick responses in the small group work, engaging in group learning, undertaking group projects and assignments have been reported as effective strategic experiences. For some there were courses where students were in teams which allowed for more in depth discussion of content (Dra.S.9), while others have experienced "opportunities to collaborate not just within their course but with other institutions" (Isa.S.11). The responses pointed to improved communication skills and engaging with the content at a deeper level as outcomes of this strategy. (Joe.A.3)

The third of the effective practices that was required is assessment (n = 2; 14.30%). In this case they believed that homework assignments and work on assignments is an effective strategic experience that they had during their courses (See Table 5.12 for more information). For instance one of the epidemiology students "...found the various assessments the lectures offered as well as the homework assignments helpful. In some courses, the multiple attempts at quizzes have been helpful in reviewing materials." (Gha.S.12) "There is a systematic method to evaluating the quality of learning and also for student assessment. This method is based on research projects, discussion group projects and student peer review." (Hel.S.6) Another stated that "I believe the assessment methods in all my classes were very successful and they were based on paper essay and worksheets and also online discussions quizzes." (Roc.S.5)

5.4.3.3 The Results of Comparative Investigation

As has been reported above, the second item of the pedagogical approach that was required was effective learning practice. Table 5.13 compares three themes that were considered in both countries. According to this table, the first priority of both countries in relation to effective learning experience required refers to online tools and activities. This result illustrates that this theme was frequently mentioned in both countries in approximately the same proportion (higher than 40%). This means

Table 5.13 Comparative result of effective learning practice required

Effective learning practice required	Percentage of comments		Comparison based on	
	AUS P	USA P	Require	Priority
Online tools and activities	43.75	42.85	Similar	Similar
Team-work	31.25	42.85	Similar	Different
Assessment	25.00	14.30	Similar	Similar

that even in America, which has been defined as the best and most progressive country in terms of technology and application of online tools, this theme is still considered to be the future of e-learning programs. Also the issue of team working has been reported by both countries in which America (42.85%) provided a higher percentage in this theme than Australia (31.25%).

This shows that the value of learning by working with others may be understood as a more effective learning practice in US rather than Australia. The third effective learning practice was reported as assessment. However, this issue was nominated more by Australians (25%) than Americans (14.30%).

5.4.4 Learning Assessment Methods Required

5.4.4.1 Australia Case Study

Within the area of e-learning assessment, the results are directly derived from the content analysis about e-learning assessment methods and reveal that there are four main themes in Australia. These are: assignments, group thread discussion, case based project and self-assessment. Table 5.14 illustrates these results.

Based on these findings, the research found that the first priority of participants' regarding assessment method for their e-learning was assignments (n = 11; 35.48%). From this perspective, assessment methods like individual assignments, individual essays, and peer review of work, reports assessment and assessment tasks have been reported as a priority by students as well as lecturers and administrative staff. For example one of the lecturers stated, "…..individual assignment format and practical activities" (JuS.F.1).

The second priority concerning assessment methods was group thread discussion (n = 8; 25.80%). In this case, as a health system associate professor explained, "We have to make assessment methods a meaningful tool based in reality and attractive in online teaching, by using contributions to discussions and reflection" (KaR.F.2).

The main comments were about different types of discussions within the online system that includes online discussions, contracts, threaded discussions, and online threaded discussion groups. See Table 5.14 for more information.

The third priority of participants was case based projects (n = 7, 22.58) which refers to case study and research projects. Comments in the interviews include case study assessment, video case study, case research, practical project, research project, case study quizzes. Senior lecturers explained that: "…blended and fully online teaching in undergraduate social psychology and undergraduate and postgraduate research methods are based on case-based research" (RoB.F.8). "Previously I used a quality improvement-QI-case study assessment method which two to three students work on together in my classes" (MaR.F.3).

Table 5.14 Assessment method required in one faculty in an Australian university

Theme	Comments of participants	Source[a]	Percent
Assignments	Individual assignments format	11	35.48
	Individual reflection essays		
	Written assignments		
	Essays and peer review of work		
	Individual essay		
	Essays and reports assessment		
	Reports and individual assignment		
	Reports and individual assignments		
	Individual work		
	Assessment tasks		
	Reports and individual assignments		
Group thread discussions	Discussions group	8	25.80
	Discussions		
	Online discussions contracts		
	Group discussion work		
	Online discussions		
	Threaded discussions		
	Online discussion group		
	Online threaded discussions group		
Case based-project	QI case study assessment	7	22.58
	Video case studies		
	Case-based research		
	Practical projects		
	Research project		
	Case study quizzes		
	Project and case studies		
Self-assessment	Self-assessment activities	5	16.12
	Self-evaluation tools		
	Self-assessment quizzes		
	Self-assessment questions		
	Self-feedback		

[a]N = 31

The last priority of the participant's responses about assessment methods was self-assessment strategies (n = 5; 16.12%). Activities like self-assessment, quizzes, questions and supplying self-evaluation tools or self-feedback have been reported by participants as possibly leading to higher levels of self-confidence and satisfaction in students. Some typical explanations from discussions were: "Not just online assessment, I believe individual essays which I think can find better results" (PaU.F.7) "In the period of a semester I have practical projects, essays and reports assessment in my online teaching" (RoD.F.9).

5.4.4.2 USA Case Study

According to American participants' results, the third area of pedagogical issues that concerned them was the e-learning assessment method. Based on participants' comments, four main e-learning assessment methods were categorized; assignments, group thread discussion, case based projects and self-assessment. Table 5.15 highlights the language which interviewees used to describe these assessment methods.

Table 5.15 Assessment method required in one faculty in a USA University

Theme	Reference	Source[a]	Percent
Group thread discussions	Group discussions and threaded discussions	11	40.74
	Group learning activities		
	Online threaded discussions		
	Group discussion projects		
	Online discussions quizzes		
	Discussions group		
	Group chatting on study		
	Thread discussions and Group presentations		
	Threaded discussions or digital media submissions		
	Discussion board quizzes		
	The group collaboration using "The Island"		
Assignments	Reports and writing	10	37.03
	Personal writing assignments		
	Homework-reports and essays		
	Paper assignments		
	Paper essay and worksheets		
	Research writing assignment and reports		
	Written out essay formats		
	Individual presentations with peer reviews		
	Write a final reflection paper or report		
	Homework submissions		
	Papers and presentation		
Case based-project	1	4	14.81
	Laboratory projects		
	Research projects		
	Projects and exams		
	Learning lab activities and projects		
Self-assessment	Self-assessments	2	7.40
	Self-assessment feedback		

[a]N = 27

The first assessment method is group discussion thread (n = 11; 40.74%). Within this assessment method, online group learning activities, discussions, projects, quizzes, and presentations have been reported by participants such as on line courses that, "…required threaded discussions or digital media submissions that have group chatting on it" (Isa.S.11).

There are challenges within this method of assessment.

There are issues in regards to feelings of frustration brought about by this on-line group involvement. One participant had the following experiences and thoughts about the inclusion of this assessment method: "I have found the group thread discussions most challenging. While the group collaboration using "The Island" was effective, many of my other group experiences have just been frustrating. I feel that group assignments should be used judiciously so as to decrease the frustration factor for students. A lot of time is wasted just trying to organize groups. I am currently taking a course that requires weekly group collaboration on simple homework assignments and I see no pedagogical value in this model." (Cyn.F.13)

As can be seen in this table, assignment (n = 10; 37.03%) was reported as a common assessment method, including reports, home works, essays and worksheets, research writing, presentations. One of the comments noted "Two kinds of assignments in the online classes required: paper assignments and lab assignments." (Hea.S.7) as well as another participant comment which focused on writing tasks saying: "My online classes have mostly consisted of a lot of writing tasks. Write weekly discussion posts, and writing a final reflection paper or report on a certain topic. Some have required that I create a document with a group. I once had an online class that presented case studies in the lecture, but very rarely did we work on one for an assignment. That may have been helpful. I've never done a video projector podcast." (Han.S.8)

The third assessment method that was reported by participants was case-based projects (n = 4; 14.81%). Doing laboratory research projects, activities and exams have been mentioned in this category, for instance a public health science lecturer explained, "We have projects which we call themes lab, which is problem solving. They have laboratory projects multipoint choice quizzes and short answers." (Kri.F.2)

The fourth and the last assessment method was self-assessment (n = 2; 7.40%). In this case, two people felt positively towards self-assessment. An e-learning instructor and faculty member highlighted this as "one of the best assessment methods for my online teaching is self-assessments and reports that are required." (Nat.F.1) Another said "the self-assessment feedback system is being updated and it is a supportive method that is necessary for students' learning improvement." (Joe.A.3)

5.4.4.3 The Results of Comparative Investigation

The third pedagogical issue in both countries was the assessment method they required. As can be seen in Table 5.16, it seems that in both countries there are similarities on the themes of this issue. Both countries claimed their issues and challenges in the assessment method were group thread discussions, assignments,

Table 5.16 Comparative result based on assessment method required

Assessment method required	Percentage of comments		Comparison based on	
	AUS P	USA P	Require	Priority
Group thread discussions	25.8	40.7	Similar	Different
Assignments	35.5	37.1	Similar	Different
Case based-project	22.6	14.8	Similar	Similar
Self-assessment	16.1	7.4	Similar	Similar

and case based project and self-assessment. Although the priority of Americans was first on group thread discussion (40.7%) and second on assignment (37.1%), the priority of Australians was first on assignments (35.5%) and second on group thread discussion (25.8%). As was shown in the last section this result indicated that learning through group activity in terms of assessment in more highlighted by Americans.

5.4.5 Effective Learning Content Required

5.4.5.1 Australia Case Study

Table 5.17 records the results of the themes of learning contents required by Australians. According to the findings, three main themes were used by the interviewees to express the content and the format that they require in their e-learning systems. These are: multimedia, online modules and texts. The results indicated that applying multimedia such as modern new formats is more attractive than online modules and texts which are regarded as traditional formats of learning by participants.

As can be seen, supplying multimedia (n = 7; 36.84%) has been reported as the first priority. They described videos such as clips and case studies, 3-D and multimedia as a priority for e-learning contents. Using multimedia in e-learning has been favoured because it facilitates learning in both simple and comprehensive ways. For example some typical explanations from discussions were: "That's a hard question! I've not evaluated it, but I would think that the most effective content is where I give students a brief visual video of an assessment and they complete an assessment report" (JuS.F.1).

> There are a range of modules to answer this question. I attempt to use visible features and animation techniques such as diagrams, charts, 3-D and in multimedia (MaR.F.3).

> Firstly I want to say all types of content in the online units are more interesting, compared with the traditional text-centred content contained in the units' handbook. If we are looking for effective content in the e-learning I think the best of them is multimedia as non-textual contents (AnK.S.12).

Table 5.17 Effective learning content required in one Faculty in an Australia University

Theme	Comments of participants	Source[a]	Percent
Multimedia	Brief visual video	7	36.84
	3-D and in multimedia		
	Multi media		
	Video case studies		
	Videos		
	Video clips		
	Multimedia		
Online module	Blackboard content	6	31.57
	Blackboard learning system		
	E-learning module		
	Virtual projects		
	Discussion boards		
	E-research module content		
Text	Text	6	31.57
	Texts to encourage thinking		
	Written texts		
	Written course guides		
	Written texts		
	Textbooks		

[a]N = 19

The second priority of participants was online modules (n = 6; 31.57%). Different types of e-learning modules like "the blackboard learning system is most effective and required in my online teaching, in that it enhances a learning environment, by providing content management and sharing" (MeR.F.4); "a range of formats seems to be the key in my e-learning module in the health sciences faculty" (MiC.F.5) and "online access to readings and e-research module content are more required in the online course programs" (SuS.S.16) have all been put forward by participants.

The third priority of content format indicated by participants was text (n = 6; 31.57%). Traditional text formats such as text to encourage thinking, written course guides and textbooks were emphasized in this theme. For example one of the participants, a senior lecturer, said "There are a variety of materials in this area. I think a mixture of text and video seems to work best for my unit and they are much more effective" (RoB.F.8). Another lecturer pointed out the use of written course guides for "…directing students to the most relevant readings, which are updated regularly" (MeR.F.4). Two of the postgraduate health science participants explained that: "written texts content is much more required than others. I believe that it is most effective and clearer than other content and is an easy way to organize your exam" (GuN.S.13). "We don't receive any 2D and 3D simulations and virtual worlds, or dynamic computer models, just referral to journal articles and textbooks" (LeA.S.14).

5.4.5.2 USA Case Study

The last area within the research of pedagogical issues relating to e-learning was e-learning content. Three main themes emerged from the study from America. These are multimedia, online modules and text. However, one of the interviewees commented that "the type of content, effectiveness and requirement of virtual classes is based on lectures." (Hea.S.7)

As can be seen in Table 5.18, multimedia (n = 9; 60%) had the highest frequency from the American study. Recorded lectures, video based lectures, video and audio presentations and multimedia lectures have been mentioned by participants. The responses have highlighted that multimedia is more effective and reinforces content more than content delivered in text only. Responses such as: "I have found multimedia lectures more helpful than simply reading PDFs. Interactive assignments using dynamic computer models have also been helpful. The use of real-world examples pertinent to my field of study has also been effective and they are recommended." (Cyn.F.13). In terms of reinforcement, responses include: "I have experience with multimedia lectures that is much more effective" (Roc.S.5) and "the audio and video presentation is most helpful with reinforcing the content" (Yin.S.10).

Table 5.18 Effective learning content required in one faculty in a USA University

Theme	Reference	Source[a]	Percent
Multimedia	Recorded lectures	9	60.00
	Multimedia case lecturers		
	Multimedia cases		
	Video-based lectures		
	Multimedia lectures and case studies		
	Multimedia lectures		
	Multimedia lectures		
	The audio and video presentation		
	Multimedia lectures and dynamic computer models		
Online module	E-Lectures module	3	20.00
	Visual e-learning module for easy-to-use		
	Moodle content		
Text	Written texts	3	20.00
	The realistic texts		
	Realistic texts		

[a]N = 15

The second content that was reported by participants was the use of online modules (n = 3; 20%). E-lecture modules and visual modules and moodle content were mentioned. For example, an e-learning designer highlighted that E-Lecture modules can be delivered from a variety of platforms (Joe.A.3). "Students have stated that they like variety" (Han.S.8) and "lectures and factual material with attractive visual e-learning modules for easy-to-use and online design presentations have been helpful" (Gha.S.12).

The third and the last e-learning content reported by interviewees was text (n = 3; 20%). Traditional texts that can be downloaded in the e-learning educational environment have been considered. For example two of the health system students note "realistic texts" (Bri.S.4) and "the type of content that is most effective is the realistic texts" (Dra.S.9).

5.4.5.3 The Results of Comparative Investigation

Comparing the results of e-learning contents in both countries, one faculty in an Australian university and one faculty in a US University showed that three themes, application of multimedia, online module and the text, are effective and required in both countries. As can be seen in Table 5.19 these results indicated that the first priority of Australians in this case is supplying online modules (31.57%) and text (31.57%), however, the first priority of Americans is in relation to multimedia content. This may be due to the fact that application of online modules and text as e-learning contents is fairly common in America rather than Australia.

5.4.6 *Cultural Issues*

To investigate the current cultural status in the Australian and American samples, participants were asked three main questions about the existence of cultural issues and sensitivities, nonexistence of cultural issues and sensitivities and effective cultural practices required. Table 5.20 summarizes the responses.

Table 5.19 Comparative result based on effective learning content required

Effective content required	Percentage of comments		Comparison based on	
	AUS P	USA P	Require	Priority
Multimedia	36.84	60	Similar	Similar
Online module	31.57	20	Similar	Similar
Text	31.57	20	Similar	Similar

Table 5.20 Participant's comments on cultural issues

Interview questions focused on	Participants	Theme	No of participant
The existence of cultural issues and Sensitive	AUS	Lack of international context awareness	13 (81.25)
		Communication issues	
		New technological Challenges	
	USA	Lack of international context awareness	10 (77)
		Communication issues	
		Differences in attitude	
The nonexistence of cultural issues and Sensitive	AUS	Similar content and quality	3 (18.75)
	USA	Popular Pandemic	3 (23)
Effective cultural practice required	AUS	Chat room group	16
	USA	Facility and renovating e-environment	13
		Improving training and skills	

As can be seen in this table, thirteen participants of one faculty in an Australian university believed that there are cultural issues and sensitivities in specific situations. They claim that due to lack of international context awareness, communication issues and the use of new technology, challenges exist between cultures. However, three people believed that there are no cultural issues and sensitivities. They believed that there is similar content and quality between cultures and there is shared popular material. Also they explained that there are effective cultural practices that can be supplied in e-learning like chat rooms, to facilitate and renovate online environments and improve training and e-learning skills.

According to the American responses as tabled, ten interviewees' comments indicate that cultural issues exist. They nominated the lack of international context awareness, communication issues and differences in attitude as the main reasons for these cultural issues.

However, three comments indicated that some believe that there are no cultural issues. They believe that similar content and quality of e-learning cultures ensures that there are no cultural issues in the e-learning system in America.

Finally thirteen people explained that effective cultural practice includes chat room groups, a renovated e-environment, and improved training and skills.

5.4.7 The Existence of Cultural Issues and Sensitivities

5.4.7.1 Australia Case Study

As has been mentioned above, there are three main reasons, namely lack of international context awareness, experiencing communication issues and new technological challenges that have been identified to account for the existence of cultural issues by Australian interviewees (n = 13; 81.25%).

As can be seen in Table 5.21, participants believed that online courses are like being in an international class in which people of different cultural backgrounds attend. The diversity of attendees should cause there to be more sensitivity about cultural customs. They claimed that builders of a standard e-learning system should be aware of the possibility of differing cultural and social norms so that these cultural issues can be addressed. In this case, some of the comments (n = 7, 35.00%) indicate there is a lack of international context awareness, for example one of the lecturers said; "I think in general there is probably some cultural awareness and cultural sensitivity. We do fully online learning; we do a couple of residentials so in those residentials we sometimes have some Indonesian students and a couple of other students from some other countries. I think in the residential, I think probably more cultural awareness there is together in the group and also I think in the international and Indonesian students I think there are some sensitive things on the technology, team work and or speaking." (RoD.F.9) Also he stated, "…total online learning lacks awareness to a large extent of cultural mores. As you know, because our University has a multiethnic and multicultural background in international settings" (RoD.F.9). One of the postgraduates in this part of the interview focused on awareness and connection with the students and explained that: "I believe most students and instructors care about improvement while posting and writing, online discussion, cafe etc. But I remember that we have had very learned instructors in the past, but they have not been able to make a connection with the students because the teachers didn't have a good knowledge of students' cultural settings—Considering the serious lack of international awareness—in order to develop or modify instruction" (SoN.S.15).

Following on from this, the participants focused on communication within the issue of cultural sensitivities (n = 5; 25.00%). In this case, there are verbal communication issues like having a language other than English as their first language. A representative comment of lecturers was: "I think about the importance of teaching being about not being exclusionary. There are some aspects of the online environment that get more feedback from some students because English is not their first language. Online discussion and the online environment has aspects of teaching that might invite students to participate more particularly if they are not confident in English and participate in the discussion rather than spend more time thinking about speaking correctly. On the other side it is no different than face to face teaching except that it is difficult to express emotions online so there is a greater chance of misunderstanding jokes for example" (JuS.F.1).

Table 5.21 The existence of cultural issues and Sensitive in one faculty in an Australian university

Yes/no	Reason	Comments of participants	Source	Percent
Yes	Lack of international context awareness	Students from overseas	7	35.00
		Students of international backgrounds		
		Multicultural background in international settings		
		International/overseas based students		
		Different cultural backgrounds and country		
		International and Indonesian students		
		Lack of international awareness		
	Communication issues	Human behaviour communication	5	25.00
		English is not their first language		
		English as second language		
		Verbal and communication skills		
		Language communicate		
	New technological challenges	Concern with new apps	4	20.00
		Advancing IT usage		
		Technological changes and feeling stressed		
		Social-networking issues		
	Total of reason		16	
Total of Yes			13	81.25
No	Similar content and quality	Same qualitative health and health systems	3	15.00
		Theoretical content		
		platform and quality are similar		
	Popular Pandemic	Global populations	1	5.00
	Total of reason		4	
Total of No			3	18.75

The third reason that has been highlighted as an issue within cultural sensitivities in relation to the e-learning system was challenges in technology (n = 4; 20%). Table 5.10 shows these results. From this perspective, they indicated that the fast speed of technological changes, new and advanced IT usage affects the e-learning system as well as the process of learning and these should be considered in relation to cultural issues. A representative comment of students was: "I just think that the e-learning and social networking are advancing very fast—new IT, advancing IT usage—and this requires educational improvement and we need to adapt to change and develop a reliable connection and also a constructive link with the social-cultural context" (AnK.S.12).

Also skills in relation to electronic communication and virtual discussion are an issue for people who have to have interaction with other students and lecturers. For example one of the lecturers said, "I found a lot of electronic virtual issues because both the lecturers and learners have sensitivities and each is capable of monitoring the network learning activity. For development of online teaching environment I think that the students and young people prefer doing their tasks, homework or meeting by searching, virtual and electronic communication and sharing a classroom" (RoB.F.8).

In contrast, three people indicated that there is no existence of cultural issues (18.75%). They put forward two main reasons for their comments: similarity of content and the quality of e-learning courses across the world which is the same in all cultures like similar qualitative health and health systems, same theoretical content and same platform and quality (n = 3; 15%). Also the popularity of online courses in an international environment based on accessing the internet and different social networks ensured that there would be no sensitivities to be aware of in regard to cultural issues (n = 1; 5%). See Table 5.21 for more information.

Some of the comments by participants illustrate both points. An Associate Professor of health systems explained: "I don't believe that the development of online teaching displays cultural sensitivity, because my blended and fully online teaching experiences are both on the same qualitative health and health systems with clear written syllabus. Also a global population is another reason that we don't have displays to show cultural sensitivity for online system development" (KaR.F.2). A science e-learning instructor commented on this: "No cultural sensitivities because these are largely centred around how we deal personally with people. Online learning is largely concerned with theoretical content, and which is not "softened" by personal interactions. Also, not being in contact with students, you are often not aware of what cultural sensitivities are required, because you are not aware of their cultural backgrounds and also the role of cultural issues is based on qualitative behavior" (MaR.F.3). An online student explained: "All students access similar platforms and quality. I am not aware of how this is differentiated or how this has been made different to accommodate cultural sensitivities and issues" (MiM.F.6).

5.4.7.2 USA Case Study

As explained in the last section, 77% believe that cultural issues exist in the American sample. See Table 5.22 for more details. In this case, three comments (23.1%) indicate there is a lack of international context awareness, for example one of the lecturers said: "I have invitational students in the course but we are going to be adding more students from different parts of the world and I almost want to say that I would like to have my course assisted by someone else to answer the

Table 5.22 The existence of cultural issues and sensitive in one faculty in a USA University

Yes/no	Reason	Comments of participants	Source	Percent
Yes	Lack of international context awareness	Different educational styles	3	23.1
		Students' instructional context		
		Pedagogical background		
	Communication issues	English as second Language	4	30.8
		Foreign language		
		Language		
		English is a second language		
	Differences in kind of attitude and perspective	Different point of view	3	23.1
		Non constructive interactive		
		New students thinking with system adaptation		
	Total of reason		10	77
Total of Yes			10	
No	Similar content and quality	All learning and materials was not insensitive per se	2	15.4
		Very natural in cultural sensitivity with congruent quality		
	Popular Pandemic	Pandemic of online environments	1	7.6
	Total of reason		3	23
Total of No			3	

culturally sensitive issues. We are concerned about creating an atmosphere of respect for all of the students." (Kri.F.2)

Taking differences into account when designing courses may help to reduce these issues. "Our University has an international position for providing online courses for students from other countries with pedagogical backgrounds particularly from the Middle East and China (Gha.S.12). Some have even noted that when courses take into account these differences that there is still the possibility of gaps in learning experiences and knowledge. "We offer many international programs specifically geared toward students from Asia, those programs curricula are specifically for those audiences but there are gaps between students' instructional contexts and e-learning units that university provided for international applicants" (Joe.A.3).

According to the results, the different educational backgrounds of international audiences is the other main reason for the existence of cultural issues in the online learning environment. "It is still following a curriculum standard that is common in the United States while we have students from other cultures and different learning styles with completely different curriculum standards. We need to move to make a global multicultural online curriculum for heath science" (Dra.S.9).

As shown in Table 5.22, communication issues have some main sub-reasons which include English as a second language with 4 comments (30.8%). International students whose first language is not English find that this can hinder their full participation in the course. As Chen states "In course designs that rely heavily on student participation for learning, 2nd language students often participate less, due to concerns about dragging a group behind, or appearing less competent" (Chen et al. 2006, p. 23). The e-learning students focused on "designing online courses" based on a second language and "development of online learning environments". As native English speakers they understand what is going on, but are "not completely sure the content is as accessible to those for whom English is a second language. The school of public health must be considerate when designing online courses and content for English as a second language for international learners" (Cyn.F.13). Others felt that their online learning environment displays cultural sensitivity. "One of them is English as a foreign and second language for international students" (Roc.S.5).

Three comments (23.07%) from students of Environmental, Epidemiology and Community Health sciences explained differences in attitude issues are like having different points of view and non-constructive interactions attest to the existence of cultural issues (Hea.S.7). There is concern that "differences in attitudes and perspectives to programs can lead to big mistakes in communication. In fact, expectations and attitudes about syllabus, labs and online activities make non-constructive or constructive interaction between students with themselves, lecturers and contents" (Han.S.8). Another felt that "there are some problems for system adaptation with new students and administrative staff thinking. However, I don't know how these will be solved but certainly contrasts in opinion is a critical and sensitive issue to discuss" (Hel.S.6).

In contrast, three people (23.09%) claimed nonexistence of cultural issues because the similar content and quality of learning materials rendered it universally culturally acceptable, and the contents are very neutral in cultural sensitivity and have congruence with all cultures (n = 2, 15.39%). For example one of the participants, an e-learning student believed that "the online learning environment would score "neutral" for displaying cultural sensitivity, meaning that all learning and materials were not insensitive per se, but also did not go out of the way to cater to other cultures" (Isa.S.11). In addition, one comment suggested the global popularity of the online environment as evidence for the nonexistence of cultural issues (n = 1, 7.70%). As Table 5.22 shows, the participant highlighted this: "The online courses I have taken thus far are very natural and popular in cultural sensitivity".

"The materials provide a diverse background of cultural, ethnic, racial, and geographical variation. We learned about the different populations when learning the concepts" (Yin.S.10). On the other hand one of the e-learning lecturers believes that "Online classes are tailored for mainstream Americans" with design of courses that is suitable for the American environment. Indeed if the university has plans for development of online programs for international purposes "there is need for more training on what the cultural sensitivity needs would be, to reflect the global population of online students" (Nat.F.1). Also one of the participants, a master of strategic communication in public health, stated "However, the university still follows a typical base for e-learning programs but they have to think about international needs and international connectivity" (Dra.S.9).

5.4.7.3 The Results of Comparative Investigation

Comparing the results of both countries, Australia and America, indicated that there are similarities and differences in their beliefs about cultural issues and sensitivity. As can be seen in Table 5.23, both countries agreed that lack of international context awareness and communication issues are two main reasons for the existence of cultural issues and sensitivity. Regardless of this similarity between the two countries, there are some differences about the main reasons for the existence of cultural issues and sensitivity. Interestingly, the theme of the challenge of new technological changes was reported just by Australians whereas the theme of differences in kinds of attitude was reported just by Americans.

In addition, two main reasons were given by both target participants as explanations for the non-existence of cultural issues and sensitivity. However, the frequency results in both countries showed that not many people believed in the existence of cultural issues and sensitivity. According to Table 5.24, the same proportion of people (15%) in both countries claimed that content and quality were similar in their e-learning systems which demonstrates the non-existence of cultural issues. In addition the second reason for non-existence of cultural issues in both countries was explained as its global popularity.

Table 5.23 Comparative result based on the existence of cultural issues

Main reason	Percentage of comments		Comparison based on	
	AUS P	USA P	Reason	Priority
Lack of international context awareness	35	23.1	Similar	Different
Communication issues	25	30.8	Similar	Different
New technological challenges	20	0	Different	Different
Differences in kind of attitude	0	23.1	Different	Different

Table 5.24 Comparative result based on the nonexistence of cultural issues

Main reason	Percentage of comments		Comparison based on	
	AUS P	USA P	Reason	Priority
Similar content and quality	15	15.39	Similar	Similar
Popular Pandemic	5	7.70	Similar	Similar

5.4.8 *Effective Cultural Practice Required*

5.4.8.1 Australia Case Study

The last issue to be investigated was effective cultural practices. In this case, participants of one faculty in an Australian university indicated that there are three effective cultural practices, namely chat room groups (n = 10; 58.82%), facilitating and renovating the environment (n = 6; 35.29%) and improving training and skills (n = 1; 5.88%).

According to Table 5.25, the first effective cultural practice indicated by interviewees was chat room groups. Virtual multicultural chat rooms which provide discussion groups in both formal and informal settings have been reported by participants as effective practices to help people understand cultural issues, concerns and problems. Table 5.25 explains all comments in detail.

One of the participants, a coordinator, believed that "administrative staff gain insights into new enrollees at the very beginning of their enquiries". Circulating student profiles to teachers and asking students at the very beginning to post introductions in class (with an emphasis on what experiences they bring to the group) are important starting points. "Teachers are exposed to continuing

Table 5.25 Effective cultural practice required in one faculty in an Australian university

Theme	Reference	Source[a]	Percent
Chat room group	Online tutorial group	10	58.82
	Virtual tool chat rooms,		
	Online multicultural chat rooms		
	Online meeting via team work		
	Online discussions		
	Informal online settings		
	Review meeting groups		
	Discussion meeting		
	Skype groups		
	Meeting and discussions group		
Facility and renovating e-environment	Flexible new services facility	6	35.29
	Update specific rules for more support		
	Renovate virtual environments		
	Online facilitators		
	Develop social networking applications and facilities		
	Facilities that support the use of Web CT		
Improving training and skills	Training improvement plan	1 5.88	

[a]N = 17

development and review meeting groups after each teaching semester to explore all areas of difficulty and for improvement, and cultural issues often form a part of these" (JaN.A.10).

Also one of the online course students explained that they "…think most talking and meeting probably makes cultural communication sometimes a little bit easier because you know you don't have any problem with accent, pronunciation and those kinds of difficulties. Also most effective I think Skype groups with friends, classmates and or with other forums" (LeA.S.14).

In addition, one of the lecturers mentioned the following about chat room and tutorial groups, "I think certainly the small kind of online tutorial group is good for this purpose and better and more effective because you can see that it encourages contributing and facilitates participation. In some small tutorial classroom environments it might be easier to identify and encourage students to participate and make quality contributions. Ask students to follow guidelines about online communication and netiquette while posting messages" (JuS.F.1).

The second effective cultural practice identified by participants was an e-learning environment that had been renovated and was facilitated.

From this perspective, comments like flexible new services facility, update specific rules for more support, renovate virtual environments, provide online facilitators, develop social networking applications and facilities, facilities that support the use of Web CT were given by participants. In regards to cultural practices, they believed that if the virtual system that users use to engage with each other to learn has been facilitated by administrators, this can simplify the process of learning.

One of the participants, a senior lecturer, believed that virtual environments and structures need renovating and explained that, "When we think about cultural communication strategies in the total online environment, we have to know and renovate virtual environments and problems to show the cultural sensitivity in the structure, in the software and hard wares, which I want to say, is effective strategy for cultural communication infrastructure" (RoD.F.9).

One of the administrative staff focused on students' encouragement saying, "we request that online facilitators encourage students to bring examples to the online classroom from their own cultural situation/country of origin or where they are living now. We also ask facilitators to provide examples from different parts of the world so that we are not focusing on one particular part of the world, and to provide a more inclusive environment. Plans are under way to develop a module in cultural competence around Indigenous communities and this is likely to be expanded to include cultural competence in general, and we would ask online facilitators to complete this" (MeG.A.11). Also a postgraduate of health science discussed new terms of communication and believed that "The University unfortunately always focused on the traditional approaches" (AnK.S.12).

The third effective cultural practice that was indicated was improving training and skills. In this case improvement of training, educational plans and skills could help people's awareness of cultural issues and concerns in students and lecturers as well as administrative staff.

One of the lecturers believed in human resources development for effective practice and said: "We need a parallel strategy; a training improvement plan and also optimally update specific rules based on human resources development for more support of cultural communication by university" (MiC.F.5).

5.4.8.2 USA Case Study

Table 5.26 displays the effective cultural practices in one faculty in a US University. As can be seen in this table, the most frequently cited effective cultural practice was the improvement of training and skills (n = 7, 52.84%). Indeed, comments like "human resources development and training, sharing and learning one's cultural background competence, education and diversity awareness activities and experiences, training to develop cultural awareness, improving cultural competence like self-awareness and other-awareness, learning styles and skills and intentional cultural communication skills and awareness" have all been discussed.

The majority of comments discuss the need to "develop cultural awareness" for effective cultural practice. Comments include:

> The focus of cultural communication strategies should be on education and diversity awareness activities and experiences and I think these strategies are effective for the online learning environment (Hea.S.7).

> In my opinion it is improving cultural competence like self-awareness and other-awareness which can be achieved through events like international conferences, workshops and seminars (Roc.S.5).

Table 5.26 Effective cultural practice required in one faculty in a USA University

Theme	Reference	Source[a]	Percent
Improving training and skills	Human resources development and training	7	53.84
	Sharing and learning one's cultural background competence		
	Educations and diversity awareness activities and experiences		
	Training, to develop cultural awareness		
	Improving cultural competence like self-awareness and other-awareness		
	Learning styles and skills		
	Intentional cultural communication skills and awareness		
Chat room group	Meetings-discussion groups	4	30.76
	Asking questions and discussions		
	Forums, chat rooms and discussions		
	Needs to have some element of real-time interaction		
Facility and renovating e-environment	Increased accessibility options	2	15.38
	Provide tools, facility and examples		

[a]N = 13

> For cultural communication I think effective strategy is training, to develop cultural awareness for students, faculty members and administration, and could involve sharing some experiences through blogs or sites or workshops (Hel.S.6).

> Online environment is a rich culture of diverse experiences. By incorporating these diversities into real world issues with human resources development and training, everyone gains knowledge, exposure and awareness of common challenges and solutions as well as areas where there are differences (Nat.F.1).

> Effective cultural communication strategies are sharing and training one's cultural background competence to bring awareness to the group and classroom (Dra.S.9).

Also four people explained that chat room groups like "discussion groups, asking questions and discussions forums are effective cultural practice" (30.76%). For more explanation, one of the participants, an online teaching instructor, believed that "meetings and chatting with students is best strategy for cultural communication" and she explains that she uses "the example where I say to students take the principle you have learned in this class and choose one major topic and explain it in the context of your field. I could adapt that to explain some health problem relevant to your culture or your community" (Kri.F.2). Also one of the participants said "The open ended nature of many forums, chat rooms and discussions are helpful for bringing out the cultural communication strategies in the online environment" (Gha.S.12). One of the participants believes that "cultural communication needs to have some element of real-time interaction, whether through webinar or chat format." (Cyn.F.13)

In addition, two comments explained facilitating and renovating the e-environment like "increased accessibility options" and "tools and facility" as effective cultural practice (15.38%). For example a health science e-learning instructional designer commented on this: "Increased accessibility options such as alternative formats like full transcripts for recorded materials; willingness to explore potential cultural gaps in programming with every new development cycle or new target audience repurposing" (Joe.A.3). Also one of the online students explained: "I think it is important to provide examples and guides within the classroom that can apply to a wide multitude of cultures, and also cater to a diversity of learning styles and skills that are dependent to existing tools and or creating new facilities" (Isa.S.11).

5.4.8.3 The Results of Comparative Investigation

Comparing the results of effective cultural practice required by participants in both countries showed that there are similarities in the expression of themes including chat room groups, facilitating and renovating the e-environment and improving training and skills in relation to this issue. However, the priority of each theme in the two countries is different. According to Table 5.27, the first priority of

Table 5.27 Comparative result based on effective cultural practice required

Cultural practice required	Percentage of comments		Comparison based on	
	AUS P	USA P	Require	Priority
Chat room group	58.82	30.76	Similar	Different
Facility and renovating e-environment	35.29	15.38	Similar	Different
Improving training and skills	5.88	53.84	Similar	Different

Australians was improving chat room groups, however, the first priority of Americans was improving training and skills. The second priority of Australians was facilitating and renovating the e-environment but the second priority of Americans was improving chat rooms. The last priory of Australians was improving training and skills however, the last priority of Americans was facilitating and renovating e-environment. This indicates that the Americans focus more on the content (train and improving skills) but the Australians focus on structures of their e-learning system.

5.5 Technological Issues

In response to the questions asked about current issues of technology in the Australian and American e-learning environment three main themes emerged. Content analysis showed that participants highlighted technological infrastructure in the Australian interviews (n = 14; 25%), which was also the case in the American interviews (n = 11; 29.72%) and the functionality of the Australian e-learning system (n = 16; 28.6%), which was also the case regarding the American e-learning system (n = 13; 35.13%). Also the results indicated that the key technological challenges were lack of training and development, old or clunky infrastructure, lack of pedagogy-driven design, poor service help-desk and networking and bandwidth issues in participants of one faculty in an Australian university responses (n = 26; 46.42%). Technological challenges including lack of pedagogy driven design, poor service from the help desk, lack of updates and lack of training and development were also indicated in participants of one faculty in an American university responses (n = 13; 35.13%). Table 5.28 displays the results.

5.5.1 Current Status of Technology

5.5.1.1 Australia Case Study

As has been mentioned above, the first technology issue reported by participants of one faculty in an Australian university was technological infrastructure (n = 14, 25%). The researcher categorized the answers of participants about this issue in

Table 5.28 Participant's comments on technological status

Interview questions focused on	Participants	Theme	No of percentage
Current status of technological infrastructure	AUS	High level	14 (25.0)
		Moderate	
		Low level	
	USA	High level	11 (29.72)
		Moderate	
Current status of functionality of e-learning system	AUS	High level	16 (28.6)
		Moderate	
	USA	Low level	13 (35.13)
Key technological challenges	AUS	Lack of training and development	26 (46.42)
		Old/clunky infrastructure	
		The lack of Pedagogy-driven design	
		Poor service help-desk	
		Networking and bandwidth issue	
	USA	Lack of training and development	13 (35.13)
		The lack of Pedagogy-driven design	
		Poor service help-desk	
		Lack of update	

[a]TI N = 14

three levels of high, moderate and low. According to results, it was found that 2 people rated technological infrastructure as good or very good (14.28%). For example Paul said "I think that IT facilities and technical infrastructure is adequate. Web pages design and helpdesk are cool. Faculty of health sciences technology has always stood out in our University for its capability to combine a standard level of network innovation with flexibility, simplicity and efficiency".

Five participants explained the technology as moderate or adequate or average (35.71%). Also seven people (50%) indicated current technological infrastructures as low, old, clunky and behind the times (See Table 5.29). To conclude, most participants believed that the current technological infrastructure is moderate and low. For example, one of the lecturers explained that "although the university may have the latest blackboard system, I am not sure that the staff has any idea of the possibilities that exist. What hampers this is the lack of new infrastructure particularly portable devices that would enable staff to explore the options outside of their time in their office or at their desk. Wireless access on this campus is very patchy

Table 5.29 Current status of technology in one faculty in an Australian university

Theme	Level	Reference	Source[a]	Percent
Technological infrastructure	High level	Good base technology	2	14.28
		The level of technology and IT services are very good		
	Moderate	IT facilities and technical infrastructure is adequate.	5	35.71
		Foundation of System is Moderate		
		Hardware System building is not at a high level		
		Upgrading IT systems is on the process now		
		Average level of technical structure		
	Low level	From 1990-ish	7	50.00
		Very old-fashioned		
		Faculty has to improve systems and Infrastructure		
		Network		
		The lack of new infrastructure		
		The system is clunky and old-fashioned		
		Quite basic and traditional system		
		Sydney system is a bit behind the times		
Functionality of e-learning system	High level	The latest blackboard system	2	12.5
		The LMS and system is a very useful specialty network and my unikey account		
	Moderate	Adequate	5	31.25
		Web pages design and helpdesk are cool		
		A standard level of network innovation with flexibility		
		Lab platforms and online standards is not at a high level		
		A project running with e-learning at the moment		
		The blackboard is a little bit difficult with tools and online app		
	Low level	Fake and no longer useful for projects	9	56.25
		Crashes or is not accessible		
		Many internet wireless black-spots		
		Is not enough standard		
		Fluid media and some contents are scary		
		Online design technology used is not standard		
		The learning lab project and online material quality is in the low level		
		Level of quality and content of design is weak		
		The online environment (quizzes. Etc.)Still is at the low level of normal		

[a]FESN = 16

for staff, and therefore limits portability. I see within one discipline the LMS is being used in a very simple manner" (MeR.F.4).

As can be seen, current technological infrastructures are in need of more improvement and "it is not of a high enough standard for faculty and as well the faculty has to improve systems and Infrastructure Network (IN)" (MiC.F.5). "Upgrading IT systems are on the process now for better technical and educational performance. I think that we could definitely improve this, and I have a project running with e-learning at the moment to do exactly that. The site being created will showcase various technologies available to online teachers, to facilitate the process of them incorporating the new approaches into their teaching. At present we are using mainly asynchronous discussion as this allows maximum flexibility for the students" (MeG.A.11).

The same categories were used to record results of participants of one faculty in an Australian university in relation to the functionality of the e-learning system. As has been shown in Table 4.13, just two people reported the functionality of their e-learning system as high (12.5%). As an online postgraduate health science student explained: "Faculty has a good base. The LMS and system is a very useful specialty network and my UniKey account has a variety of options like eLearning, email, my Uni and Student software" (GuN.S.13).

With five people indicating functionality as moderate (31.25%) and nine (56%) as low, most of the interviewees believe that the system has a low level of functionality. Explanations from the interviews include:

> I think our system is a bit behind the times. From memory, it's far behind my previoUS University, which I attended from 2000–2003. That university even then offered pod cast lectures, online tutorials, and online assignment submission for most courses I entered. This university isn't really there even now in my experience and the online environment (quizzes etc.) still is at the low level of normal (SuS.S.16).

> We use the online blackboard learning system; I think that the blackboard is a little difficult with tools and online app sometimes especially the Wait setup. It could probably function a little better. They could set up real time systems for presentations, webinars, videos or other additional technology that would help especially with complicated topics (LeA.S.14).

> There is a gap between the existing state and the ideal state of affairs and some parts of them need to improve. Level of quality and content of design is weak. We need the pedagogical design studio based on a design-centred framework (SoN.S.15).

> Quite basic and traditional system frankly, the learning lab project and online material quality is in the low level. There is not good balance between the university brand and the level of technology that exists (AnK.S.12).

5.5.1.2 USA Case Study

To investigate current technological status the researcher asked two main questions of participants of one faculty in an American university about technological infrastructure and functionality of e-learning system. As with the Australian case study the researcher categorized the responses into three levels of high, moderate and low. Table 5.30 shows the results.

The results of this content analysis revealed that most of the participants of one faculty in an American university evaluated technological infrastructure at moderate (n = 6; 54.54%) or high (n = 5; 45.45%) levels. It is interesting that none of them felt that the technological infrastructure warranted a low level response.

As can be seen in Table 5.30 the positive responses show that the participants for the most part feel that there are excellent facilities that work well and that the educational technology application is “very accessible and easy to use” (Hel.S.6). The infrastructure is seen to be a system that “is a robust one” (Gha.S.12).

Responses within the moderate category included comments that expressed systems that work for the needs of the students despite not always being the most up to date. Participants can see that more technology could be incorporated but budget constraints aside “it is effective for LMS” (Isa.S.11) and that technological modesty “allows us to troubleshoot nearly all student issues” (Joe.A.3).

Results show that the participants evaluated the level of functionality of the e-learning system in the following categories: high (n = 7; 53.85%), moderate (n = 5; 38.46%) and low level (n = 1; 7.69%). It was felt that there are “excellent resources; materials; applications; good online instructions and tutorials” by those who evaluated functionality as high. One of the lecturers with more than six online teaching experience explained; “I think we have excellent resourses and superbly trained staff because the people who are in the digital learning group are in structural design and are web designers. They work very effectively with us and that support is fantastic”.

> According to the system that has been improved professors are using a greater variety of tools to teach the information (videos, audio clips, lectures in print and audio, links to different websites). I'm not sure what you could change but I'd be open to trying new things (Kri.F.2).

Within the moderate response to functionality, comments were “need for routine updates, the current level of online system is adequate, nothing attractive about the current level, it works for the purposes and needs” (Nat.F.1). There is acknowledgement that there is room for improvement alongside the evaluation of an e-learning system that is functional. (Bri.S.4). Some of the health science e-learning instructors and students highlighted this: “I think it works for the purposes and needs of the class, but I don't think it's innovative or using the highest technology available through the university” (Roc.S.5).

Only one participant evaluated the functionality of the e-learning system at the low level. As Table 4.27 shows the participant highlighted the following: “Functionality of the online program is sub-optimal. The quality of the tutorials for statistical software was poor and difficult to listen to. Timed exam format on

Table 5.30 Current status of technology in one faculty in a USA University

Theme	Level	Reference	Source[a]	Percent
Technological infrastructure	High level	Excellent facility with the base tools	5	45.45
		It works well		
		Very strong		
		The level of technological infrastructure is good		
		There is robust system being used		
	Moderate	Level of technology is modest	6	54.54
		The current level of technology has improved		
		More technology could be incorporated		
		Normal but nothing outstanding		
		Moderate infrastructure		
		Has been reasonable for school		
	Low level		0	
Functionality of e-learning system	High level	Excellent e-learning resourses	7	53.85
		The educational technology application is very accessible and easy to use		
		Output of e-learning system is very satisfactory		
		Good online instructions and tutorial system		
		Level used is effective for teaching the material		
		The online pages are well organized and excellent		
		Using a greater variety of tools to teach		
	Moderate	Need for routine updates	5	38.46
		The current level of online system is adequate		
		Nothing attractive about the current level		
		It works for the purposes and needs of the class		
		Modesty		
	Low level	Functionality of online program is sub-optimal	1	7.69

[a]TI N = 11
[b]FESN = 13

Moodle was extremely frustrating to use because navigating the site was difficult. In my current class the web links from the lecture slides frequently don't work" (Cyn. F.13).

5.5.1.3 The Results of Comparative Investigation

Comparing the results of Australians and Americans about the current status of the technology in their e-learning revealed that there are significant differences between the perception of participants in the two countries about technological infrastructure. As can be seen in Table 5.31 participants of one faculty in an American university reported the level of technological infrastructure on their e-learning system as high or moderate. From this point of view none of the Americans evaluated this issue as low level. On the other hand the majority of Australians believed that the current status of technological infrastructure is low or moderate and only a minority of them believed that the level of technology in the Australian e-learning system is high.

The same pattern can be seen in the evaluation of Americans and Australians about functionality of the e-learning system in both countries. As can be seen in Table 5.32 the majority of Americans reported that the functionality of the e-learning system is high or moderate however, the majority of Australians reported this factor as low or moderate.

5.5.2 *Key Technological Challenges of E-Learning*

5.5.2.1 Australia Case Study

The second question to enquire into current technological issues was about current e-learning challenges. In this case five main themes were highlighted from Australian participants' responses namely lack of training and development (n = 8; 30.76%), old or clunky infrastructure (n = 7; 26.92%), lack of pedagogy driven

Table 5.31 Comparative result based on technological infrastructure

Technological infrastructure	Percentage of comments		Comparison based on	
	AUS P	USA P	Level	Priority
High level	14.28	45.45	Different	Different
Moderate	35.71	54.54	Different	Different
Low level	50.00	0	Different	Different

Table 5.32 Comparative result based on functionality of e-learning system

Functionality of e-learning system	Percentage of comments		Comparison based on	
	AUS	USA	Level	Priority
High level	12.5	53.85	Different	Different
Moderate	31.25	38.46	Similar	Similar
Low level	56.25	7.69	Different	Different

design (n = 5, 19.23%), poor service help-desk (n = 3; 11.53) and networking and band width issues (n = 3; 11.53%). Table 5.33 shows the results. For example one of the participants claimed several challenges and said "I have found out that there is some sort of challenge with technology, infrastructure and networking services. There was a quiet system some time ago that made interrupts on my courses. The other one is that the IT infrastructure was not sufficient to engage with the faculty. One more thing is the lack of plans as to how to use technology for staff and students in the faculty. I had experience previously that the design of technology is based on the technology driven design approach and it is not based on the pedagogy-driven design for online learning environment" (JuS.F.1).

The first challenge indicated by participants was lack of training and development of educational programs and courses. From this perspective it has been shown that lecturers and administrative staff do not have enough knowledge to improve their skills and abilities to supply, develop and progress the educational e-learning system. Two of the main comments of participants were:

> …faculty should set up a short course to improve the online and technical skills and ensure basic training for new and inexperienced staff and also new students; alongside lack of innovation and update with new practices and skills as well as clearly linking pedagogy with e-learning practice (KaR.F.2).

> Resources in terms of staff who could be given more time to become familiar with the technologies and then facilitate the process of the technologies being adopted in various units of study (MeG.A.11).

The second challenge that technological issues face in the e-learning system is old infrastructure. In this case, comments like "IT infrastructure was not sufficient to engage with the faculty; lack of innovation and update with new practices and system; reliability of equipment and existing infrastructure; the reliability of Blackboard infrastructure and the Blackboard is clunky and slow to use" have been articulated. A lecturer of occupational therapy stated, "I can't believe I've come to a university with the reputation it has. To be tied to my desk without any new system with a big HD, a small monitor, no laptop, no tablets or iPods or other portable devices; and no easy access to good videoconferencing facilities on my campus, working with colleagues who still bring pen and paper to their meetings. My previous academic employer was much smaller, with a much smaller overall budget, yet technologically is far better equipped to meet the challenges of educating the 21st century learner" (MeR.F.4). Also one of the postgraduate students

Table 5.33 Key technological challenges in one faculty in an Australian university

Theme	Reference	Source[a]	Percent
Lack of training and development	The lack of plans as to how to use technology	8	30.76
	Short course to improve the online and technical skills		
	Knowing how to use some of the e-learning elements		
	Time to learn and staff development		
	Training staff		
	Lack of knowledge about how and what online learning can offer		
	Training and development skill of new technology		
	Familiar with the technologies and then facilitate the process		
Old/clunky infrastructure	The IT infrastructure was not sufficient to engage with the faculty	7	26.92
	Lack of innovation and update with new practices and system		
	Reliability of equipment and existing Infrastructure		
	Without any new system with a big HD		
	The reliability of Blackboard infrastructure		
	Blackboard is clunky and slow to use.		
	Breakdown of Technical Infrastructure		
The lack of Pedagogy-driven design	It is not based on the Pedagogy-driven design	5	19.23
	Linking pedagogy with e-learning practice		
	Develop quality of learning		
	The content design of instruction in Blackboard		
	Lack of presence of the design of the module-integrated		
Poor service help-desk	ICT service reply and help-desk feedback on problems	3	11.53
	No facilities and supports		
	Help desk services		
Networking and bandwidth issue	No easy access to good videoconferencing facilities	3	11.53
	Bandwidth and wireless issues for remote locations		
	Security access to the online courses and LMS Limitations		

[a]N = 26

stated, "Always I have big challenges with the e-learning system. A key challenge is Blackboard. It is clunky and slow to use" (AnK.S.12).

The third challenge indicated by respondents was the lack of pedagogy driven design. Specifically, participants claimed that the focus of their e-learning educational system is more on technology rather than pedagogy. Indeed, emphasis on attractive design format and visual content rather than the quality of pedagogy has been mentioned as an issue. For example one of the participants said: "I had experience previously that the design of technology is based on the technology driven design approach and it is not based on the pedagogy-driven design for online learning environment" (JuS.F.1).

A lecturer of an online health course believes that it is time that the focus was on the quality of pedagogy and explained that it is "time to develop quality of learning and pedagogical design", and that this "is the biggest challenge. It takes more effort and time to develop quality online learning experiences. To do it well requires time" (MiC.F.5).

A student of online learning with experience in five courses claimed that there is a lack of presence within the design of e-learning modules which could lead to students being put off from using such technology (SoN.S.15).

A poor service help desk was the fourth technological challenge that was indicated. In this case participants believed that there are not enough facilities to support suppliers in virtual systems to help them, to ask questions or to solve problems. For example two of the main comments of participants were: "....ICT service reply and help-desk feedback on problems which is very poor and sometimes wrong" (AnK.S.12). Another said that there were "no facilities and supports for real time communication and educational research. Because when we have complicated topics and questions and when we have to write down the questions and then wait two days for someone to write the answer back, it doesn't work" (LeA.S.14).

The last challenge shown by participants was poor networking and low band width issues which is simply about the weak management of the e-learning system. One of the participants, an administrator who supports students, said, "We also have to be aware of bandwidth and wireless issues for remote locations" (JaN.A.10) and one of the postgraduates in health sciences sees issues with security regarding access to the online courses and LMS: "In my opinion the key challenge is security access to the online courses and LMS limitations, to prevent people who are not enrolled in Units of Study from accessing Units of Study discussions and assessments" (GuN.S.13).

5.5.2.2 USA Case Study

Research into the key technological challenges of e-learning in USA revealed four main themes: a lack of pedagogy driven design, poor service help desk, lack of updates and lack of training and development. Table 5.34 shows these results

Glossy technologies without attention to solid pedagogical purpose, inflexibility within the design of LMS, which highlights technology more than pedagogy, have

Table 5.34 Key technological challenges in one faculty in a USA University

Theme	Reference	Source[a]	Percent
The lack of Pedagogy-driven design	Glossy technologies and forget that faculty must serve a solid pedagogical purpose	4	30.76
	Design of LMS is not instructional flexible enough		
	Technology is very highlighted and curriculum content and instructional quality is poor		
	Focused on using flashy technology than on learning the material		
Poor service help-desk	The non-support of certain major browsers by Moodle	4	30.76
	Not monitoring and support discussion boards or not replying to comments mand any help		
	Help desk technological support is one of the main issues in public health school		
	Without additional contact with the professor or teaching assistant		
lack of update	Tool of the day is not updated with new application versions	3	23.07
	Links and app not working or not updated		
Lack of training and development	Faculty who don't understand online teaching and it's difficult to dispel the methods about teaching	2	15.38
	Administrative staff and help desk services who don't understand something about e-learning systems		

[a]N = 13

been mentioned as elements of the lack of pedagogy driven design (n = 4; 30.76%). While it is great to have the latest technologies, budgets do not stretch to cover production costs and do not consider return on investment in regards to the number of students who may use these technologies (Joe.A.3). Even with all of these factors being considered often there is little thought to whether a technology serves the content, with transfer of knowledge being the objective (Isa.S.11).

In terms of support of the e-learner, the research highlighted that help from the service help desk is poor (n = 4; 30.76%) with a need for technological support required. "E-learning practice is not helped when some platforms are not supported by some major browsers" (Yin.S.10) was one comment along with "the lack of updates to content to ensure that it is relevant" (Hel.S.6).

It can be seen from comments made by participants that "lecturers need to monitor and respond to questions and concerns of their students" (Han.S.8). Others explained: "From my experience so far it's more the individual instructor's use of the technology that's problematic".

> Web links don't function and it is far too common for typos to obscure content also links and app not working or not updated. Presentations are not updated from one semester to another, rendering assignments unclear (Cyn.F.13).

Participants of one faculty in an American university also indicated that e-learning system tools, applications and URL links are not up to date (n = 3; 23.07%). Sometimes lecturers fit the students to the tool they wish to use rather than using tools that will effectively communicate the course content, with one participant noting his particular difficulty: "Deciding what is the "tool of the day" to meet all the students when our students represent such a wide diversity in age, geographical location and access to technology" (Nat.F.1).

Training and development is the last theme within technological challenges highlighted in the American research. It seems that the faculty environment like services, tools and utilities doesn't protect online programs and that administrators need to learn more about help-desk services and the e-learning system (n = 2; 15.38%).

Regarding challenges of lack of training, one of the lecturers believes that; "there are two key technological challenges: (1) It is hard when we change platforms. It's a lot of work to shift to new platforms without any training, like converting from WebCT to Moodle. (2) More than that I found challenges to communicate to faculty who don't understand online teaching and it's difficult to dispel the methods about teaching" (Kri.F.2).

Also one of the online students added: "One of the challenges is about administrative staff and help desk services who don't understand something about e-learning systems" (Hea.S.7). Surprisingly there were no comments about old or clunky infrastructure, networking and band width issue.

5.5.2.3 The Results of Comparative Investigation

There are similarities and differences between Australia and America in terms of key technological challenges. As can be seen in Table 5.35, the similarities of both countries were lack of training, lack of pedagogy-driven design, and poor service help desk. However, the primary key technological challenge reported by Australians was lack of training and development of e-learning system, whereas the primary key technological challenges in America were lack of pedagogy driven design and poor service help-desk. The second key technological challenge for Australians was the presence of old or clunky infrastructure however, this issue was not reported by Americans. As has been shown previously this may be because the infrastructure of e-learning in America is more advanced. In contrast, the second

Table 5.35 Comparative result Based on key technological challenges

Key technological challenges	Percentage of comments		Comparison based on	
	AUS P	USA P	Challenge	Priority
Lack of training and development	30.76	15.38	Similar	Different
Old/clunky infrastructure	26.92	0	Different	Different
The lack of Pedagogy-driven design	19.23	30.76	Similar	Different
Poor service help-desk	11.53	30.76	Similar	Different
Networking and bandwidth issue	11.53	0	Different	Different
lack of update	0	23.07	Different	Different

key technological challenge in America was lack of updating the e-learning system. This theme was not mentioned by Australians. The third key technological challenges in Australia were poor service help desk and networking and bandwidth issue equally. Interestingly, weakness of networking and bandwidth was not mentioned by Americans. The last key technological challenge in America was lack of training and development, which was the first key technological challenge for Australia.

5.5.3 *Best Aspects of the E-Learning Environment and Areas that Need Improvement*

The last issue that the investigation focused on was the best aspects of e-learning and the challenges faced which, if addressed, could lead to an improved system. To obtain this information two main questions were asked: what are the best aspects of e-learning and what needs to improve in the e-learning system. Table 5.36 shows the results.

As is illustrated in Table 5.36, four main themes emerged from the responses of participants of one faculty in an Australian university to the first question. Flexibility, collaborative problem solving, accessibility and self-efficacy environment were all mentioned by interviewees as the best aspects of their e-learning system (n = 26; 50.98%).

The Australian answers to the second question have been categorized into five main themes. The need for improvement covered faculty policies and procedures and the quality of materials that are used in e-learning system in regards to pedagogy. LMS design, implementation and empowerment of staff and online support assistance emerged as areas that need review and change in the technological side of the e-learning system (25; 49.02%).

Table 5.36 Participant's comments on best aspects and need improvement

Interview questions focused on	Participants	Theme	No of percentage
Best aspects of e-learning	AUS	Flexibility	26 (50.98)
		Collaborative problem solving	
		Accessibility	
		Self-efficacy environment	
	USA	Flexibility	18 (54.55)
		Accessibility	
		Self-efficacy environment	
		Minimum costs	
Need improvement of e-learning	AUS	Faculty policies and procedures	25 (49.02)
		The quality of materials used	
		LMS design and implementation	
		Empowerment of staff	
		Online support assistance	
	USA	Faculty policies and procedures	15 (45.45)
		The quality of materials used	
		LMS design and implementation	
		Online support assistance	

On the other hand in the USA sample as can be seen in Table 4.29, the best aspects of e-learning were flexibility, accessibility, self-efficacy environment and minimum costs (n = 18; 54.55%). For example one of the participants noted several factors as best aspects of her e-learning program and highlighted this: "I appreciate the flexibility of being able to work full time and still continue my education. I also appreciate the diversity of my fellow students. In each of my classes, students have participated and been accessed from across the country and around the world, and I feel that enriches the educational experience" (Cyn.F.13).

In terms of areas that need attention research shows that faculty policies and procedures, the quality of materials, LMS design and implementation and online support assistance need improvement (n = 15; 45.45%).

5.5.4 *Best Aspects of E-Learning*

5.5.4.1 Australia Case Study

As can be seen in Table 5.37, the best aspect of e-learning with the highest frequency of comments belonged to the flexibility of the e-learning educational system in Australia (n = 10; 38.46%). In this case flexibility of time, location, courses that people are interested to attend and learning opportunities for both students and academic staffs have been indicated by participants. JuS.F.1 said "Students don't need to be on campus; they can work at their own convenient time. One of things that I found was e-learning from the first year was moderated flexibly by the few working academics in these kinds of situation, which was the best aspect of the online program". Also one of the admins of e-learning explained that "flexibility for postgraduate full-time professionals with high level of teacher/peer interaction maximizes learning opportunities and retention" (JaN.A.10).

In addition to this a student with the health e-learning system claimed to "not having to waste time and money getting to and from class" and explained that "there is flexibility around time of attendance with online classes, ability to be travelling far from Sydney while still engaging in class and also the ability to learn about the experience of others in faraway places because they also do not need to be in situ to 'attend'" (SuS.S.16).

The second aspect that has been defined in terms of best practice was collaborative problem solving (n = 7; 26.92%).

It was mentioned that the collaborative environment of e-learning encourages people to share their ideas and findings which in turn leads to progress in the science field. Two of the senior lecturers noted: "… The other thing is practical application in the health sciences is based on problem solving with engaging which is solution-oriented" (MaR.F.3) and "it is an approach for sharing and finding new ideas" (JuS.F.1).

According to comments "the students are finally being encouraged to engage and work together in the LMS" and a collaborative environment is "created online whereby students can discuss issues amongst themselves with a content expert facilitating the discussion" (MeR.F.4) and also "sharing experience and ideas are a fantastic aspect of studying with online units, that allows you to be precisely who you need to be in that moment" (SoN.S.15).

Following on from this, the third best aspect of e-learning has been defined as accessibility of e-learning sources (n = 5; 19.23%). Specifically the participants believed that easy and quick accessibility to resources and materials as well as conversations between students, lecturers and administrative staff including webinars between lecturers and students while they are overseas is one of the great aspects in an e-learning educational system. Typical explanations from discussions stretch from: "A range of learning resources available for students and staff are the best aspects of an online program" (MiC.F.5) to "the best aspect is easy access to important material such as texts" (Rod.F.9).

Table 5.37 Best aspects of e-learning in one faculty in an Australian university

Theme	Reference	Source[a]	Percent
Flexibility	Moderated flexibility by the few working academics	10	38.46
	There is flexibility for staff and students		
	The flexible time		
	Providing learning opportunities to people who can't attend campus		
	Flexibility for postgraduate full-time professionals		
	The high levels of flexibility		
	Flexibility in time and space is highlighted		
	Flexibility of study and the subjects learned based on interest		
	Flexible delivery		
	There is flexibility around time of attendance with online classes		
Collaborative problem solving	Very collaborative approach for sharing and finding new idea	7	26.92
	Practical application in the health sciences is based on problem solving with engaging.		
	The students are finally being encouraged to engage and work together in the LMS		
	Available tools and a supportive system for students that they work together.		
	Improved communication and collaboration with students		
	A collaborative environment		
	Sharing experience and ideas are a fantastic aspect of studying		
Accessibility	Capacity to integrate and access a range of materials	5	19.23
	Learning resources available for students and staff		
	Immediately viewable and am able to discuss with students in forums.		
	Easy access to important material		
	Access online resources, online conversation and webinar		
Self-efficacy environment	Communications, cognitive socially and present on focus of self.	4	15.38
	Self-paced and organised multimedia		
	Own schedules with high self-efficacy		
	Self-efficacy and organizing		

[a]N = 26

The last aspect that participants felt should be included in best practices of e-learning environments is self-efficacy (n = 4; 15.38%).

Improving self-efficacy via the process of learning in the e-learning system was highlighted by participants. "Letting people set their own schedules with high self-efficacy. Not having to give boring big lectures" (RoB.F.8) to "in my online teaching programs I think the best aspects are self-paced and organized multimedia" (PaU.F.7).

According to this, some of the participants believed that engaging with the e-learning system helped them to improve their self-confidence in regards to selecting tools for their learning as well as self-management, organization and engagement in the learning process.

5.5.4.2 USA Case Study

According to American participants' results as has been shown in Table 5.38, the first best aspect of e-learning was flexibility (n = 10; 55.50%).

Flexibility in the learning environment and flexibility in regards to time within the e-learning environment has been mentioned by interviewees. Based on the significance of flexibility in the e-learning environment for the interviewees, we can see that it is important as "flexibility in the learning and teaching process (including materials, resourses, assessments and methods) shrinks the world to a more knowable, flexible and understandable size" and so "different skills and learning styles can be utilized" (Kri.F.2).

Another noted that they had "online courses previously, rather than this online semester so I want to say that it is very flexible environment for teamwork and assignments" (Hel.S.6).

In addition some of the participants focused on flexibility in regards to time with responses from research highlighting the elimination of travel time (Hea.S.7) and ability to be studying from any location (Han.S.8), along with how a student can schedule their own time around work and family commitments.

> Flexibility with regards to my other commitments - I can work on assignments on weekends when I don't have work. It allows me to work at my own pace..." (Yin.S.10). There is also the "fantastic benefit of being able to go back and listen to lectures" (Roc.S.5). "It is flexible for the professional and mature students. Online courses are an excellent way for completing course work especially on part time basis (Gha.S.12).

The second best aspect of e-learning has been explained as accessibility (n = 6; 33.33%). Easy access to online courses and materials, forums and discussion groups has been indicated. For example one of the health science e-learning designers commented on this: "Online courses provide wide access to traditional

Table 5.38 Best aspects of e-learning in one faculty in a USA University

Theme	Reference	Source[a]	Percent
Flexibility	Flexibility in the learning and teaching process	10	55.50
	Knowable and flexible" and understandable size		
	Flexibility and eliminating travel to campus,		
	Very flexible		
	Time flexibility and also being able to go back and listen to lectures		
	Flexible time		
	The flexibility for the professional and mature students		
	It's somewhat flexible		
	Flexibility		
	The flexibility of being able to work full time		
Accessibility	Greater access for students	6	33.33
	Online courses provide wide access		
	Have access to the forums 24/7 to discuss the class		
	Access the database		
	Ability to convey the material to a large audience students have been participated and accessed from across the country and around the world		
Self-efficacy environment	Solving the problem by myself without the need for instructors	1	5.55
Minimum costs	Great exposure at minimum costs	1	5.55

[a]N = 18

and non-traditional students; local and international. They also provide twenty-four seven easy access, although this raises questions about student support" (Joe.A.3).

One of the occupational therapy students with over six online course experience believes that "I am able to work when it fits into my schedule and have access to the forums 24/7 to discuss the class with other students and the instructors" (Dra.S.9). "Also there is the "ability to convey the materials to a large audience as well as instructors from across the country/world" (Isa.S.11).

The third best aspect of e-learning in American research was a self-efficacy environment (n = 1; 5.55%). Self-problem solving and constructive learning has been indicated in this case with one of the students saying "….I also like the fact that online learning allows for immediate correction and response to quiz/test questions and solving the problem by myself without the need for instructors to correct the assignment" (Isa.S.11).

The last best aspect of e-learning was minimum costs of e-learning (n = 1; 5.55%) in comparison to traditional face to face learning.

As a health science e-learning instructor commented on this: "Tuition at this university remains one of the lowest of the top-ranked schools of public health and great exposure at minimum costs with e-learning courses in Biostatistics, Epidemiology and Community Health, and Health Policy and Management and Environmental Health Sciences area.

Also the school of public health has a range from online short courses trainings to offering CEUs (Continuing Education) for free" (Nat.F.1).

5.5.4.3 The Results of Comparative Investigation

As can be seen in Table 5.39 our enquiry into the best aspects of e-learning in both countries revealed that the primary best aspect was flexibility of the e-learning system. The theme of collaborative problem solving was mentioned as the second best aspect in just Australia. Accessibility of the e-learning system was mentioned in both countries as one of the best aspects of e-learning. According to results America provides better accessibility than Australia. The next best aspect of e-learning belonged to a self-efficacy environment and it seems that Australia provided this theme better than America. The last best aspect that was only reported by Americans was minimum costs. This theme was not reported in Australia.

5.5.5 Areas that Need Improvement Within the E-Learning Environment

5.5.5.1 Australia Case Study

The last theme investigated in this research program concerned the aspects which need to improve for the future. From this perspective, according to Australian interview results there were five main themes: faculty policies and procedures

Table 5.39 The comparative result based on best aspects of e-learning

Best aspects of e-learning	Percentage of comments		Comparison based on	
	AUS P	USA P	Best aspects	Priority
Flexibility	38.46	55.50	Similar	Similar
Collaborative problem solving	26.92	0	Different	Different
Accessibility	19.23	33.33	Similar	Different
Self-efficacy environment	15.38	5.55	Similar	Different
Minimum costs	0	5.55	Different	Different

(n = 7; 28%); the quality of materials used (n = 5; 20%); LMS design and implementation (n = 5; 20%); empowerment of staff (n = 4; 16%) and online support assistance (n = 4; 16%). Table 5.40 further illustrates these results.

Concerning the first of these aspects, participants believe that faculty policies and procedures need more time invested in them in regards to new principles and strategies, and resources and courses for the online system. For example one of the

Table 5.40 Areas that need improvement within the e-learning environment in one faculty in an Australian university

Theme	Reference	Source[a]	Percent
Faculty policies and procedures	Faculty need time for new procedures	7	28.00
	Protocols and policies that look for how to make it better		
	Strategic use of new multimedia tools and lack of new procedures and rules		
	Faculty have to provide new and strong policies for Security		
	Change and modify organisational structure and process		
	Policies of increasing online courses		
	Lack of practical approaches for new educational sittings like e-research		
The quality of materials used	We need to support quality of discussion board content	5	20.00
	Linking pedagogy and quality of practice		
	Provide more effective resources and content		
	Innovative online pedagogical curriculum		
	Effective presentation of learning material		
LMS design and implementation	The design, format and some apps need to improve	5	20.00
	Technological design update		
	Issues with the user-friendliness of the LMS		
	How implementation of new-technological content and design		
	Blackboard system design		
Empowerment of staff	Human resources; ongoing staff training	4	16.00
	Training		
	Up-skill staff to achieve knowledge of e-learning		
	The poor background of teachers/facilitators		
Online support assistance	The access to the hardware/software	4	16.00
	Educational and technical online assistance		
	IT assistance is lacking in certain areas		
	Faster feedback		

[a]N = 25

administration and student coordinators stated, "we have regular meetings in every semester in the Uni where we find during these meetings some sort of problem related to learning. I think one of the most important challenges of e-learning practice is strategic use of new multimedia tools and lack of new procedures and rules" (JaN.A.10).

Issues in relation to online security have also been highlighted by a health science postgraduate who said that the "faculty have to provide new and strong policies for security to restrict access to Units of Study only to enrolled students, University administrators and facilitators" (GuN.S.13). Other comments also focused on specific policies that need improvement such as increasing the number of online courses offered: "this University could extend its e-learning programs further…to more classes to policies of increasing online courses. I had to take time out of my work day to attend two (undergraduate) units that could have easily been delivered online" (SuS.S.16).

Some of the comments were focused on the procedures that need to change and be modified for example: "The faculty needs to change procedures to play around with available tools and consider which ones will be most effective for them and the achievement of learning outcomes" (MiC.F.5). Another thought that modification of the organizational structure was needed for "more interaction plus effective presentation of learning material" (SoN.S.15).

The second need illustrated by interviewees was better quality of materials used in the e-learning educational system. It has been mentioned that quality of the content, materials and resources that have been presented as curriculum should be improved, linking pedagogy and quality of practice, and faculty needing to support the quality of discussion board content.

For example one of the administrators noted, "I think we need to be moving into how to implement new-technological content and design, which we are planning to do from next year onwards. We could provide more effective resources and content of pedagogy for those who like to approach study from different angles—so videos to better explain concepts, real-time linkups at times, to talk to fellow students and teachers and cover key concepts. Involving students even further in terms of obtaining feedback—the ITL surveys at times seem to give limited open responses" (MeG.A.11).

The third aspect that respondents felt should be improved is the technological side of e-learning, with improvements made to the LMS design and implementation. For example some typical explanations and comments reflected these issues: "The design, format and some apps and LMS need to improve" (JuS.F.1). Issues of usability have surfaced: "Usually I have issues with the user-friendliness of the LMS. E-learning and ICT should modify some systems also I think about lack of practical approaches for new educational sittings like e-research and blackboard system design" (PaU.F.7).

The fourth aspect that needs improvement is staff empowerment. Training for all staff whether it is administrative or academic in the most up to date technology has been highlighted. In this case strength of staff in both technology and pedagogy needs to improve. For example, one of the lecturers believes that something needs to improve to "ensure students develop the skills to be e-learners and ensure staff has the opportunities to up skill their knowledge of e-learning" (MeR.F.4). A health postgraduate student explained that the "faculty need educational empowerment of staff, I think of the poor background of teachers/facilitators because some of them don't have professional experience with e-learning apps and innovative online pedagogical curriculum" which impacts the professional reputation of higher education institutions (AnK.S.12).

The last aspect identified as being in need of improvement was online support assistance. Participants claimed that support in both technical and pedagogical aspects should be improved. One felt that "…creating a sense of shared learning in the students via educational and technical online assistance" was needed as "the big plus of face-to-face is students informally helping, and learning from each other. It's important to recreate the informal learning environment online" (RoB.F.8). While another commented about the lack of IT assistance in "certain areas such as lecture recording and other on line activities. More assistance with Turnitin (Internet-based plagiarism-prevention service) is needed etc." (RoD.F.9).

In conclusion we can see from the above that while responses showed that there were great aspects to e-learning in both pedagogy and the technology of the system, there are still improvements that need to be addressed for e-learning to move forward within one faculty in an Australian university.

5.5.5.2 USA Case Study

Table 5.41 displays the areas that need improvement in e-learning in the USA sample. Based on this table, the first area that needs to improve concerns faculty policies and procedures (n = 6; 40%). In this case some of the participants discuss resources and funding policies for the development of online course systems. For example one of the administrative staff, an instructional designer/developer, stated, "Online learning can be very easily scaled up, but if done right, it's very expensive to develop. There needs to be more business discussion about resources and funding policies for development of online courses. What should the average course cost in terms of staff time, etc.? What is too much? What is too little? The variables of online course development and newness of the practice don't provide a lot of good answers on this" (Joe.A.3).

Also some of the participants cited learning process and strategies relating to teachers, students and the online classroom as areas that they feel need to improve. Consistency by teachers within their subject was highlighted: "It is always more tough doing online classes because you aren't in class to ask questions, so if teachers could keep it consistent that'd be best and the school should change some roles" (Bri.S.4). There were also concerns about redundancies in the materials and sources and "assignments that are more time consuming and do not add much value to the course" (Gha.S.12). Another participant felt that "the school needs to review the strategies for more engaged teachers in touch with the virtual world inside the

Table 5.41 Areas that need improvement within the e-learning environment in one faculty in a US University

Theme	Reference	Source[a]	Percent
Faculty policies and procedures	The school need more activities for updating process, materials and system	6	40.00
	Resources and funding polices for development of online courses		
	The school should change some roles		
	Need to revision in assignments providing policies		
	E-learning strategies of school for better effectiveness		
	The school needs to review the strategies for more engaged teachers		
The quality of materials used	The quality of videos	5	33.33
	The quality of applications and also some materials of courses		
	Redundancies in the material and sources		
	Old contents		
LMS design and implementation	Modify some app process and add some tools	2	13.33
	Could include video chat tool within LMS		
Online support assistance	Support and problem solving	2	13.33
	The limited after-hours help support available		

[a]N = 15

online classroom. Professors need to take into account the real, logistical issues their students may have with collaborating and meeting goals" (Yin.S.10).

The second area that neds improvement indicated by interviewees was the quality of materials (n = 5; 33.33%). In fact some of the participants believed that the school of public health has to improve the quality of applications and materials of courses, and also the quality of the way lectures are recorded and viewed. Every teacher seems to use different software, and the quality of some of them is poor. For example some of the students' and lecturers' notes from this discussion express this: "I have concerns about the quality of videos; some of them are not so good. I think it would be better if we could use some animation programs and some 3D for online practice" (Hel.S.6). "Too much reliance on discussion boards. I realize that's how professors know you are engaging with the material, but it gets old contents!" (Han.S.8). "The group projects seem random and more designed for the ease of instructors than for the benefit of students. There needs to be a higher standard for proof reading materials and being sure that links are functional. For example, in my biostatistics class there were often misprints in equations. For a beginning student, it was impossible for me to catch this type of mistake, leading to confusion and frustration" (Cyn. F.13). "Incorporating current events in a timelier manner. Classes are prepared well ahead of sessions. So if there is a current foodborne outbreak and the class is working on a module dealing with air quality, students are not interested in looking back to food safety issues if that module has already been dealt with" (Nat.F.1).

On the other hand, some of participants concerned about quality of learning as their first priority felt that there is no clear pedagogical vision for online content during discussion. Participants impressed that the course content is just cobbled together to meet requirements. A representative comment was:

> I think that it is important to not let technology get in the way of learning. It is important that learning the material is the first priority. I also think that a somewhat unexplored option for online learning is to provide a "create your own adventure" format to learning where students could focus their attention on aspects of the class that are more difficult to them, or they find more useful. Just spitballing here, but theoretically an instructor could create a class that is broad in theory, but has multiple tracks that a student could take that would result in them learning information that is most applicable to their chosen career or area of interest (Isa.S.11).

The third area that needs improvement was LMS design and implementation like modification of the application process, adding video or chat tools to the e-learning system (n = 2; 13.33%). For example, the participants explained why they think these are necessary: "We could include video chat tools within LMS. The video chat is really useful and necessary for improving learning skills, and good understanding as well as to access communication and discussion between students (Hea. S.7). Another felt that "online learning allows me to wait until the day an assignment is due before starting any of the lectures. So we must modify some app process and add some tools and or process for online units" (Dra.S.9).

The last area indicated that needs to be improved is online support assistance. Interviewees want the quality and quantity of help support in the problem solving system to be improved (n = 2; 13.33%). For instance, one of the administrative staff said, "Online learning happens 24/7. Our office is not staffed 24/7, but often, I feel compelled to answer student, faculty and staff requests outside of office hours because these requests couldn't be handled by the limited after-hours help support available" (Joe.A.3). Also one of the health science e-learning instructors commented on needing "the opportunity for discussion in small groups between the instructor and the students for support and problem solving. We need to be able to have conversations and think and talk about the complexities of the issues and that's hard when all we are doing is reading on the discussion boards and supportive services. I lost my voice by going online and it was really good when I was given the recorded lectures. I want to be able to speak more with students and it's not so easy to right now (Kri.F.2).

5.5.5.3 The Results of Comparative Investigation

As can be seen in Table 5.42, comparing the issue of need for improvement showed that in both countries the first priority that needs to improve belonged to faculty policies and procedures. The second priority of both countries was in relation to the quality of materials that are used in the e-learning system. To continue, the third priority in both countries belonged to improvement on LMS design and

Table 5.42 The comparative result based on need improvement

Areas that need improvement	Percentage of comments		Comparison based on	
	AUS P[a]	USA P	Need to improvement	Priority
Faculty policies and procedures	28.00	40.00	Similar	Similar
The quality of materials used	20.00	33.33	Similar	Similar
LMS design and implementation	20.00	13.33	Similar	Different
Empowerment of staff	16.00	0	Different	Different
Online support assistance	16.00	13.33	Similar	Different

[a]Participants

implementation. This factor was highlighted more by Australians (20%) than Americans (13.33%). Surprisingly only Australians claimed empowerment of staff; this issue had no concern in America. The last priority of both countries was about the online support assistance; the need to improve this was higher in Australia than America.

The following chapter contains a general discussion and conclusion and looks at the key findings of the three studies. It then discusses the limitations, and makes suggestions for future studies.

Chapter 6
Conclusions

6.1 Introduction

This chapter brings together the key findings of Chaps. 3, 4 and 5 based on the aims and questions of this research. The main aim of the research study was to make a comparison of the dominant cultural dimensions, the current status of e-practices and current issues and problems between an Australian university and an American university that will ultimately improve the quality of e-learning programs in Australian and American universities. To reach this aim the researcher asked three main questions namely:

1. What are the dominant cultural dimensions of e-learning practice in an Australian and an American university?
2. What are the current status of e-learning practices in an Australian and an American university?
3. What are the dominant issues of e-learning practices in an Australian and an American university?

For each of the research questions, a summary of the research results, together with limitations and suggestions for future studies will be discussed separately in the following.

6.2 What Are the Dominant Cultural Dimensions of E-Learning Practice in an Australian and an American University?

As explained in the results, the cultural dimensions have been measured by 9 main factors and 21 questions with two orientations of instructivism and constructivism. To contribute a broader understanding of the similarities and differences of cultural

S.H. Sadeghi, *E-Learning Practice in Higher Education: A Mixed-Method Comparative Analysis*, Studies in Systems, Decision and Control 122,
https://doi.org/10.1007/978-3-319-65939-8_6

dimensions between Australian and American participants, comparative results have been reported based on their academic position in e-learning systems.

Comparative results with respect to *the educational paradigm*, which have been reported based on academic position, showed that the majority of Australian students described the system as sharply focused and towards instructivism which has been illustrated as being sharply focused on learning objectives for knowledge acquisition. However, the majority of American students described the system as unfocussed. Both Australian and American lecturers' perspectives on learning approaches were sharply focused through instructivism. Surprisingly, the majority of Australian administrative staff described the system as constructivism but the majority of American administrative staffs' perspective was toward instructivisim. Therefore, as can be seen in Table 6.1, the lecturers and students in Australia described learning objectives, path to learn and the approach of assessment, with a sharply focused learning dimension and predetermined goals rather than unfocused learning goals but the administrative staff, in contrast, described unfocused learning goals, approaches and assessment. On the other hand, the results of the participants of one faculty in an American university showed that the American lecturers and administrative staffs' ideas about the educational paradigm were towards instructivism, the same as the Australian lecturers. This result was mainly consistent with Morris' (2009), Washburn's (2012), and Gamble's (2009) results which showed the U.S. participants described the system as a structured learning plan where the goals, schedule, path and assessment were focused. However, in this study it has been found that the American students described the system as unfocused. This difference between the result in this project and other past research may be due to the fact that during the years from 2009 to 2015, the accessibility, method and procedures of e-learning have progressed. The fact that the science of e-learning in both technology and content has been enhanced may cause the perspective of American students to have changed from instructivism to constructivism.

Finally according to the results, it seems that the dominant aspect of Australian participants' descriptions was toward sharply focused on the educational paradigm. However, American participants' descriptions were placed more between the two dimensions regarding both the educational paradigm and learning approaches. This result was consistent with Edmondson's (2004) result that showed the American group described a structured learning plan where the goals, schedule, path and assessment were between a focused and an unfocused orientation. Indeed the result

Table 6.1 The dominant orientation of educational paradigm

Variables	AUS participants		USA participants		Comparison
	Sharply Focused	Unfocused	Sharply Focused	Unfocused	
Position					
Student	✓			✓	Different
Lecturer	✓		✓		Same
Admins		✓	✓		Different

of the pedagogical paradigm showed that Australian participants' description is more influenced by an eastern cultural background and traditional approaches, however, this influence in American participants' orientation is less and results pointed more toward equal description. It can be concluded that even in US which has a high quality of e-learning contents and facilities there is still a traditional perspective of instructivism towards learning.

The factor of experiential value with two orientations of abstract (instructivism perspective) and concrete (constructivism perspective) was measured by three main questions about *congruence of learning with reality, outcome orientations of learning and practicality of learning*. Overall comparative results with respect to the dimensions of experiential value reported based on academic position, showed that the majority of Australian students, lecturers and admins described the system as sharply focused and towards an instructivism approach.

However, the majority of American students and admins described the system as unfocused and toward a constructivism approach. This description is consistent with Morris' (2009) research about Asian students in the United States and Washburn's (2012) result about cultural sensitivity for western course designers in the United States. Also perspectives of American lecturers leaned more toward equal descriptions between the two dimensions. This result is consistent with Edmondson's (2004) and Gamble's (2009) results about American cultural context on experiential value. Table 6.2 shows that participants of one faculty in an Australian university clearly described learning by using contextualized examples and traditional approaches which contrasts with the American perspective which described learning by doing or from practical experience.

The role of instructor, with an indicator of didactic based on the instructivism dimension and an indicator of facilitative based on the constructivism dimension, was measured by two main questions, *one related to learning needs and one related to source of learning*. As can be seen in Table 6.3, in Australia and U.S., all participants reported that the didactic role of instructor related to both learning needs and source of learning was present in their e-learning system which is indeed the role focused on a teacher-centred approach. They claimed that students follow a path of learning determined by the instructor because the instructor usually knows what students need to learn and what source is best to learn. Comparing the results showed that there were no significant differences between the dominant aspect based on position within or without country context except for American lecturers and admins. As can be seen in the table, in U.S., lecturers believed that a

Table 6.2 The dominant orientation of experiential value

Variables	AUS participants		USA participants		Comparison
	Abstract	Concrete	Abstract	Concrete	
Position					
Student	✓			✓	Different
Lecturer	✓		Equal		Different
Admins	✓			✓	Different

student-centred and facilitative role of instructor in relation to learning needs and source of learning was present in their e-learning system. However, the American admins' descriptions of learning in their system were nearly equally divided between a teacher centred approach, and a facilitative student-centred approach. This result is consistent with Washburn's (2012), Gamble's (2009) and Edmondson's (2004) results. Also according to results, the perspectives of Australian students, lecturers and administrative staff and American students are shaped based on a teacher centred approach but the perspectives of American lecturers and administrative staff are shaped based on a student centred approach. Despite e-learning practice having developed independently in the two countries, it seems that students in both countries tend to be taught by an expert and they think that the teacher knows what they need to learn. This teacher centred description is consistent with Morris' (2009) result.

The value of errors factor has two poles of value: toward errorless learning or toward learning based on experience. This factor was measured by two main subjects namely *being errorless in the process of learning* and *the satisfaction of course designer about learning*. Table 6.4 shows that in both countries the dominant orientation of all participants based on academic position was towards satisfaction with the course designer when students learn from their mistakes and learn from experience which they explained as students learning from their errors by experimenting with what they have learned. Based on these results the dominant aspect of all Australians and Americans regardless of their positions is toward constructivism on this factor. The results are consistent with Gamble's (2009) and Edmondson's (2004) results.

The fact that 50% of Australian lecturers and admins described errorless learning as being integral to their practice may be due to the fact that getting experience in internet based courses may cause a desire for perfectionism in learning in the suppliers. Wanting to be perfect in learning may happen due to the isolated environment of learning (learning at home individually not at the university with others) but accessibility to learning materials (download e-books, journals, articles) may also shape this idea in Australian lecturers and admins. Although this assumption need to be tested in further research.

The origin of motivation with two dimensions, extrinsic and intrinsic motivation, was measured by three aspects namely *requirements of learning, reasons for learning and preference of learning*. The dimension of extrinsic motivation

Table 6.3 The dominant orientation of role of instructor

Variables	AUS participants		USA participants		Comparison
	Didactic	Facilitative	Didactic	Facilitative	
Position					
Student	✓		✓		Same
Lecturer	✓			✓	Different
Admins	✓		Equal		Different

Table 6.4 The dominant orientation of value of errors

Variables	AUS participants		USA participants		Comparison
	Errorless learning	Learning from experience	Errorless learning	Learning from experience	
Position					
Student		✓		✓	Same
Lecturer		✓		✓	Same
Admins		✓		✓	Same

reflected an instructivism orientation and the dimension of intrinsic motivation reflected a constructivism orientation.

Table 6.5 shows that in Australia the dominant opinion of all participants is that there is an extrinsic origin of motivation for requirements of learning in their e-learning system, claiming that students take e-learning courses when they are required to (to pass the course or take a degree).

However, in one faculty in a US university the majority of students and administrative staff described that there is an intrinsic origin of motivation for requirements of learning, claiming that students take e-learning courses when they want to (are genuinely interested in learning new knowledge or skills). On the other hand lecturer participants were equally divided between those who reported intrinsic and those who reported extrinsic motivation. The results indicated that the dominant aspect of Australian perspectives to origin of motivation is toward instructivism which is in contrast to the American perspective.

Therefore, the results indicated that there were significant differences between the dominant aspects in the two countries. However, the American result is not consistent with Edmondson's (2004), Morris' (2009), Gamble's (2009) and Washburn's (2012) results, because Edmondson's (2004) and Gamble's (2009) results showed that the U.S. participants' opinions were more equally distributed. On the other hand Morris' (2009) and Washburn's (2012) results did show that U.S. students are more motivated to learn from extrinsic dimension. Indeed the result of research about this factor shows that student' motivation over the past years has moved from extrinsic to intrinsic motivation.

The accommodation of individual differences factor has two poles of differences: toward non-existent or toward multifaceted. The dimension of non-existent which does not consider individual differences, reflects an instructivism orientation. The

Table 6.5 The dominant orientation of origin of motivation

Variables	AUS participants		USA participants		Comparison
	Extrinsic	Intrinsic	Extrinsic	Intrinsic	
Position					
Student	✓			✓	Different
Lecturer	✓		Equal		Different
Admins	✓			✓	Different

dimension multifaceted which considers learners' individual differences and accommodates the course to meet those preferences with metacognitive support techniques, indicates a constructivism orientation. This factor has been measured by two main variables namely *learning activities* and *consideration of needs and interests in learning*. As can be seen in Table 6.6, the results of this factor showed that all participants of one faculty in an Australian university and American lecturers believed that learning activities which can accommodate individual differences are non-existent in their system. From this perspective, mostly they claimed that the instructor or course designer uses very few learning activities and methods which allow students to learn just through predetermined methods. Also they explained that students' interests and needs are usually not considered in designing and providing courses (learning resources).

However, the American administrative staff perspective is completely contrary to the American lecturers' results which is consistent with Morris' (2009) and Washburn's (2012) results. Indeed American administrative staff described that in their system there are multifaceted learning activities which accommodate individual differences. They claimed that the instructor or course designer uses a variety of learning activities and instructional methods (like problem solving, case analysing, participation, etc.), so that students can utilize what most suits their affect. This result is consistent with Gamble's (2009) and Edmondson's (2004) results. On the other hand the perspective of American students was equally divided between those who believed the system does not consider individual differences and those who consider it does (multifaceted).

The factor of learner control has two poles: non-existent and unrestricted. The pole of non-existence of learner control reflects the orientation of instructivism and the pole of unrestricted learner control reflects the orientation of constructivism. The factor of learner control was measured by two main factors namely limitations in learning and source of learning materials. As can be seen in Table 6.7, the results of learner control in Australia and America showed that there is similarity regarding the dominant orientation of instructivism in their e-learning practice. Indeed in both countries most participants strongly believed that the deadline for learning activities was defined for students by the system and instructors and also they believed that the e-learning system defined a strong restriction for learners in both countries.

This result is consistent with Morris' (2009) and Washburn's (2012) findings but not consistent with Edmondson's (2004) and Gamble's (2009) results, because the Edmondson and Gamble results indicated nearly equal reports of learner control

Table 6.6 The dominant orientation of accommodation of individual differences

Variables	AUS participants		USA participants		Comparison
	Non-existent	Multifaceted	Non-existent	Multifaceted	
Position					
Student	✓		Equal		Different
Lecturer	✓		✓		Same
Admins	✓			✓	Different

Table 6.7 The dominant orientation of learner control

Variables	AUS participants		USA participants		Comparison
	Non-existent	Unrestricted	Non-existent	Unrestricted	
Position					
Student	✓		✓		Same
Lecturer	✓		✓		Same
Admins	✓		✓		Same

whereas Morris and Washburn results showed that the participants tended to think control was non-existent. The result appears to be reflective of a changing cultural paradigm toward teacher- and instructor-centred learning from 2004 to 2015.

The user activity factor with two dimensions, mathemagenic and generative, was measured by two aspects namely knowledge engagement and learning resources. The mathemagenic activity dimension indicates an orientation to confined learning access. Generative refers to a propensity towards easily accessible learning material. As can be seen in Table 6.8, the results of this factor showed that all participants of one faculty in an Australian university and the American students believed that in their country's practice students do not have any involvement in producing and representing knowledge which is the mathemagenic, passive approach to acquiring knowledge; also the students usually access representations of provided learning resources according to a predetermined path. However, administrative staff, lecturers and American males believed that their students are engaged in the process of creating knowledge and they create their own uses of the learning resources within the course. This perspective was completely contrary to the Australian result but it is consistent with Morris' (2009) and Washburn's (2012) results.

The factor of collaborative learning with two dimensions, unsupported and integrated, was measured by two subjects namely approach of learning activities and facilities of learning. The dimension of unsupported learning indicates the orientation of instructivism and refers to learning by oneself without collaboration; however, the dimension of integrated learning reflects the orientation of constructivism and learning through collaborative activities. Comparing the results between one faculty in an Australian university and one faculty in a US university showed that there were no significant differences between Australian and American opinions. Table 6.9 shows that in both countries the dominant orientation of participants except admins is approximately toward a constructivism approach. The American

Table 6.8 The dominant orientation of user activity

Variables	AUS participants		USA participants		Comparison
	Mathemagenic	Generative	Mathemagenic	Generative	
Position					
Student	✓		✓		Same
Lecturer	✓			✓	Different
Admins	✓			✓	Different

Table 6.9 The dominant orientation of collaborative learning

Variables	AUS participants		USA participants		Comparison
	Unsupported	Integrated	Unsupported	Integrated	
Position					
Student		✓		✓	Same
Lecturer		✓		✓	Same
Admins	Equal		✓		Different

results are not consistent with Edmondson's (2004), Morris' (2009), Gamble's (2009) and Washburn's (2012) results, because their result showed that the U.S. participants' perceptions are near the middle and they nominate the dimensions of unsupported and integrated equally.

In conclusion it is worth mentioning that according to the data of this research program, the dominant orientation of learning in both countries relies on a teacher centred perspective. Specifically, this research found that the dominant aspect of the cultural dimension in an Australian university is towards instructivism in most factors. The probable reason for this may be due to the fact that most participants who took part in this research belonged to an international community and were from undeveloped eastern countries rather than the domestic environment. According to a study of Australian higher education, the rate of international students who attend e-learning courses in Australia is higher than domestic students (DIISRTE 2013). This study also indicated that most international students who prefer to take part in online courses are from eastern countries like China, Indonesia, and Malaysia and so on. This shows that the dominant perspective of participants of one faculty in an Australian university may be affected by the cultural background from which they came (see Fig. 6.1).

In contrast, as shown in Fig. 6.2 in America the dominant aspect of learning was focused on a student centred orientation which shows the level of development of the learning environment in America in both quality and quantity. This constructivism approach to e-learning reflected the level of development and progress as well as the cultural social background of the participants especially students. While America has been known as the most progressive country in the world in relation to technology this can affect the e-learning educational perspective of its suppliers including administrative and academic staff as to the process of learning in e-learning courses.

Indeed the more novel the technological educational system, the stronger the perspective of constructivism of the participants. One approach that shows the power of the American educational e-learning system in contrast to the Australian e-learning system concerns the certificates and approved degrees resulting from e-learning courses. According to Ho et al. (2014), most students who take part in e-learning programs in US have an academic degree as a result of passing the online course (Norton 2014).

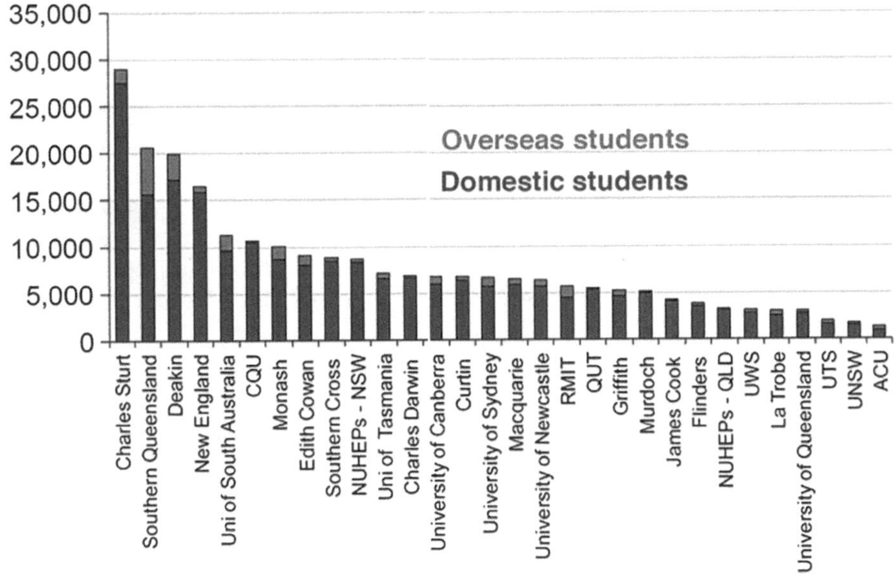

Fig. 6.1 Australian e-learning system based on University, 2011

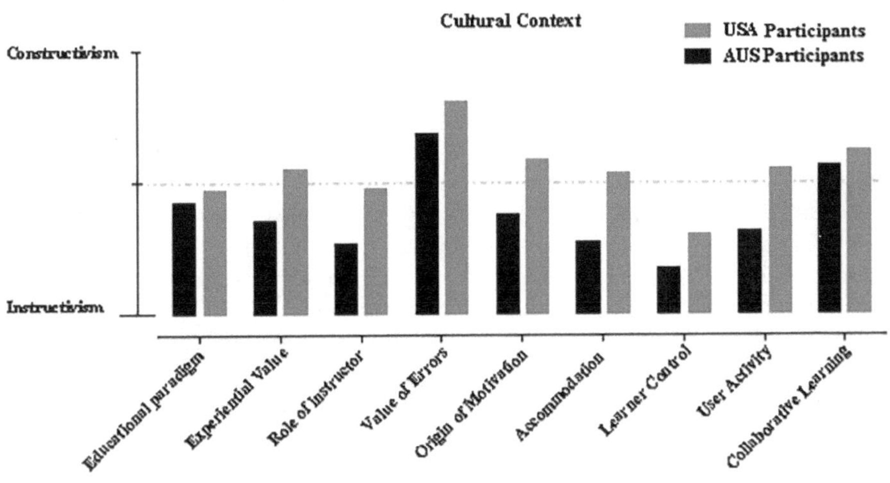

Fig. 6.2 Mean level of dominant aspect of the cultural dimension

6.2.1 Limitations

One limitation that this research struggled with was the uncertainty regarding the level of international versus domestic students, which may affect the results. To bypass this limitation it has been suggested that future research provide information about the nationality of participants to indicate whether they are international or domestic.

Another limitation of this research study was the disinclination of other higher educational institutes in Australia and America to take part in this study. Further research could provide more useful information using a range of different universities to compare the pattern of cultural dimension between them.

The first study of this book discussed the dominant cultural orientations of e-practice in one faculty in an Australian university and one faculty in a US university e-learning systems and then compared the results of the two faculties. In the next study the current status of Australian and American e-learning practices is discussed.

6.3 What Is the Current Status of E-Learning Practices in the Australian and American Universities Studied in This Book?

In this section, the key findings of the current study, limitations, implications and suggestions for further research are presented. The aim of this study was to evaluate the e-practice factor in an Australian and an American university. The findings of this research study showed that in general participants evaluated e-practice factors above average. However, the results of this study showed that the participants of one faculty in an American university evaluated e-practice factors higher than Australian participants.

The results of six main factors and each sub factor have been reported for both America and Australia for the purpose of comparison.

Regarding the evaluation of the pedagogical e-practice factor, both countries believed this was above average but American students and lecturers assessed this factor higher than Australians. However, Australian administrative staff evaluated this factor higher than American administrative staff. The higher evaluation of pedagogical e-practice by American students and lecturers indicated that e-learning programs and practice in this country are more pedagogically oriented than in Australia, although the evaluation of e-learning programs is also good in Australia. This can be seen in the perspective of Australian administrative staff who evaluated this e-practice higher than Americans. The gap that exists between the evaluation of Australian students and lecturers on one hand and Australian administrative staff on the other hand may be due to the fact that Australian administrative staff regard pedagogical e-practice less realistically than Australian students and lecturers. Staff may just consider pedagogical e-practice in terms of policy and procedure rather than considering this factor in relation to outcomes and to connections between students and lecturers. Both countries evaluated the sub factors of student centred interactivity, socio-communication, assessment and learning environment as above average. Also, learning resources and learning environment were evaluated as excellent in both countries.

In terms of each of the sub factors of pedagogical e-practice namely student centred interactivity, socio-communication, assessment, learning resources, and learning environment, American students evaluated these sub factors higher than Australian students. Furthermore American lecturers evaluated the sub factors of student centred interactivity and assessment higher than Australians. There were no differences between the evaluation of American and Australian lecturers of the other pedagogical e-practice sub factors. The fact that two sub factors were evaluated more highly by American lecturers than by Australian lecturers plus the fact that the highest evaluation of this factor and its sub factors belonged to the perspective of American students implies that pedagogical practice of e-learning in America is more fitted to the needs of learners. Australian administrative staff evaluated the sub factors of assessment and learning environment higher than Americans. In contrast, American administrative staff evaluated the sub factor of socio-communication higher than Australians. This shows that engaging socially with the e-learning system and adoption of an e-learning system with a social perspective is considered important by American administrative staff. There were no differences in evaluation of the rest of the pedagogical e-practice sub-factors between those participants.

In terms of technological practice, the results of this study showed that both the American participants and participants of one faculty in an Australian university evaluated this factor and its sub factors as above average and excellent, however, Americans generally evaluated them better than Australians. American students and lecturers evaluated technological e-practice higher than Australians; however, there were no differences between evaluations of American and Australian administrative staff of this e-practice. American students evaluated all sub factors of technological e-practice higher than Australian students and American lecturers evaluated the sub factors of technological platform, accessibility and interface design higher than Australian lecturers. There were no differences in evaluation of other sub factors between American and Australian lecturers. This implies that the levels of technological practice, LMS system and the functionality of programs in America are superior in terms of quality. The fact that the Australian e-learning system has used the technological products of the American e-learning system also reveals that the level of technological e-practice is relatively higher in America. Surprisingly, Australian administrative staff evaluated the sub factors of technological infrastructure and accessibility higher than American administrative staff. In contrast, American administrative staff evaluated the sub factors of technological platform and interface design higher than Australians. Evaluation of the reusability and technological e-practice factors was the same for staff in both samples. The results indicated that the perspective of Australian administrative staff about the e-learning system is generally optimistic. The fact that administrative staff evaluated most factors higher than the others may be because they wanted to convey the idea that they performed an important role in the e-learning system.

The evaluations of instructional design practice and its sub factors were above average in general in both countries; however, the sub factor of organizing resources was evaluated as poor in the Australian sample and poor and average in the American sample. This indicates that this sub factor needs to improve in both countries. American students evaluated the factor of instructional design e-practice higher than Australian students, however, lecturers and administrative staff evaluated this factor in both countries the same. Also, American students and lecturers evaluated the sub factor of learning scenarios higher than Australians. There were no significant differences in evaluation of the remaining sub factors between Australian and American students and lecturers. American administrative staff evaluated the sub factors of personalization, organizing resources, and accuracy materials higher than Australians; however, Australian administrative staff evaluated the sub factor of clarifying expectation higher than the Americans. Based on these results it seems that the quality of instructional design in America is higher than in Australia.

The factor of organizational practice was also evaluated in both countries as above average and excellent. The fact that staff evaluated this factor and its sub factors as excellent may be because they were more engaged with this e-practice factor.

The factor of organisational e-practice was evaluated the same by students and administrative staff of both countries, however, American lecturers evaluated this sub factor higher than Australians. Also, American students evaluated the sub factors of institutional affairs, and precedent reputation higher than Australian students. This may be due to the fact that the American university has a higher ranking in terms of facilities and institutional affairs. There were no differences between American and Australian students in evaluation of the rest of the sub factors. Also, with the exception of the sub factor of institutional affairs which was the same in both countries, American lecturers evaluated all sub factors higher than Australian lecturers. There were no differences between evaluations of all the sub factors of organisational e-practice by administrative staff of both countries.

Evaluations of support practice and its sub factors were above average and excellent in both countries, although the American staff and lecturers tended to evaluate this factor higher than others. Support practice was evaluated higher by all participants of one faculty in an American university than by Australians. Overall, the technical support sub factor was evaluated higher by American students, lecturers and administrative staff. This sub factor is generally related to the factor of technological e-practice which was evaluated higher by Americans as well and indeed, this sub factor was evaluated higher by all American participants. In addition, the academic support sub factor was also evaluated higher by American lecturers. There were no differences in evaluation of the rest of the sub factors between countries. These results imply that the level of support services for all academic positions is higher in America.

The e-practice of performance appraisal and its sub factors was also evaluated as above average and excellent; staff in both countries evaluated all these sub factors favourably. The e-practice of performance appraisal was evaluated higher by

American lecturers, whereas students and administrative staff evaluated this practice the same in both countries. In terms of the evaluation of the sub factor of performance appraisal, students of both countries were satisfied above average (see result section) about their e-learning system. The sub factor of satisfaction was evaluated by American lecturers and administrative staff higher than by Australians. Only Australian administrative staff evaluated the sub factor of cost effectiveness higher than American administrative staff, the rest of the sub factors were evaluated the same in both countries. Administrative staff in both countries also evaluated the performance of e-learning higher than average.

In general as shown in Fig. 6.3 this research study indicated that in both countries e-practices were above average. However, participants of one faculty in an American university evaluated e-practice factors higher than Australian participants. Also pedagogical, instructional design and performance appraisal practices in both countries are evaluated to some extent close while there is more gap between the two countries' participants' evaluation of technological, organisational and support service practices.

6.3.1 Limitations

This research was conducted based on a survey, however, there are other qualitative methods such as classroom observation and document analysis which could have

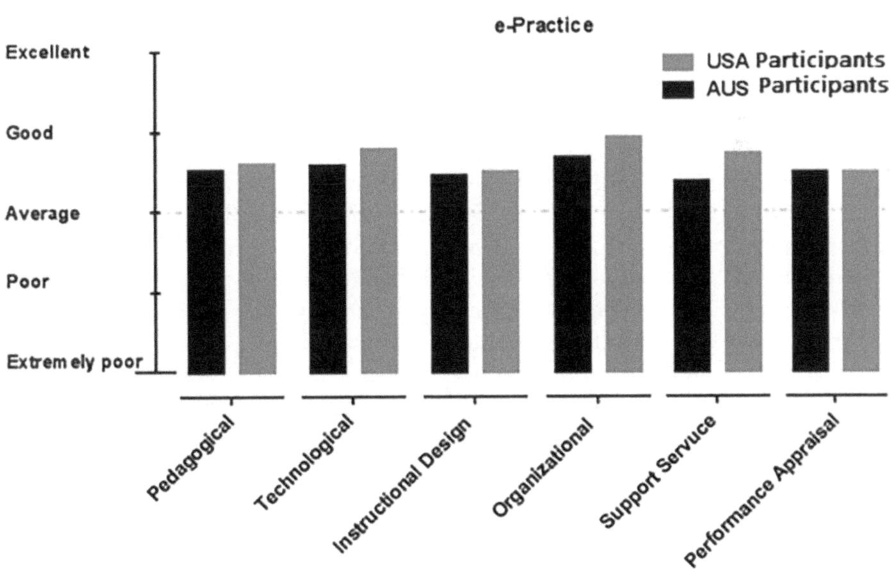

Fig. 6.3 Mean level of e-practice factors

been used to investigate policies, and review guidelines. Time and budget limitations restricted the researcher from conducting this study using these qualitative methods. Further research could be conducted based on observation and review of documents. Another limitation that the researcher encountered in this study was the limitation of a self-constructed questionnaire of e-practice. After reviewing the literature, it was apparent that a comprehensive questionnaire that evaluates all aspects of e-practice did not exist. Instead, different researchers simply explored some aspects of e-practice factors by applying different questionnaires separately. Further research should focus on conducting a comprehensive study to evaluate all e-practice factors and build some large scale e-practice components. Regarding the e-practice questionnaire that was used in this study, it could be used with a larger sample size and with different cultural background participants in order to validate it and perhaps derive some generalisable data.

The second study of the book discussed the current status of Australian and American e-learning practices. The next study posed the current problems of the two countries as the final study. Current problems are discussed in relation to the pedagogical, technological and cultural aspects.

6.4 What Are the Dominant Issues of E-Learning Practices in the Australian and American Universities Taking Part in This Study?

In this section, a summary of findings is briefly reviewed with respect to their meaning in American and Australian cultures. The research identified main items of pedagogical, cultural, technological challenges which need to improve. Furthermore, the best aspects of the e-learning systems in both countries were determined. The first finding referred to the approach of participants to learning. The results of our interviews showed that there are two main issues about the approach to learning in those countries: 1-requirements and 2-priorities. In both countries participants believed some requirements for their approach to learning were collaborative learning, problem based learning and feedback based practice. The results for this item showed that both countries tended to move towards a constructivism perspective about learning approaches. However, this preference for constructivism cannot be satisfied by the current e-learning system in both countries so the e-learning systems need to be improved and fitted to those requirements in the future. Apart from the similarities, another requirement regarding approach to learning, mentioned only by Americans, was a requirement to be outcome oriented. This means that Americans consider the progress and outcomes of learning in their e-learning system as another requirement of the approach to learning. It seems that the factor of practicality in learning is more considered by Americans rather than Australians.

The second issue in relation to requirements for approach to learning concerns priorities. The first priority participants mentioned was collaborative learning. That this priority was high in both countries illustrates the importance and significance of this item in American and Australian e-learning systems. The results also showed some differences in priorities in America and Australia. The second priority of Americans was outcome oriented learning, the third priority was problem based learning and the last priority was feedback based practice. However, in one faculty in a Australian university, the second priority belonged to feedback based practice and the third and last priority was problem based learning (Fig. 6.4).

A desire for feedback based practice is the second priority in Australian participants. This means that in this country facilities like having assistance or help desk services which can help providers to connect more with students should be promoted. However, in one faculty in a US university it seems that this requirement has already been achieved in some levels as they claim this priority as the last one.

The next theme in the field of pedagogy referred to the effectiveness of learning practice. Like the first theme, we found that there were two main issues, requirements and priorities, considered in both countries. Effective learning practices such as team working, different methods of assessment and online tools and activities have been reported as requirements in both Australia and America. The results mean that in both countries the e-learning providers know that to achieve collaborative learning they require to improve team working. Also creating and applying new and various methods to assess students as well as novel digital tools and instruments that can encourage students to learn in a team should be instituted in order to progress towards effective learning practice.

As regard to priorities, the results showed that the first need that should be considered by both countries is using novel and new online tools and activities,

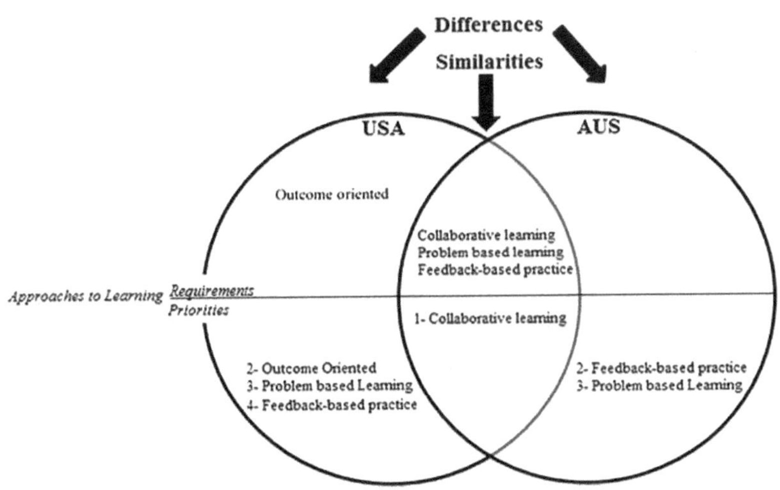

Fig. 6.4 The comparative result based on approach to learning

although in America improving team working had the same priority as the need for online tools and activities which shows that Americans have reached a comprehensive understanding that team working and using new online tools and activities would be effective when suppliers apply them together. To continue, the second priority Americans claimed was updating new methods for assessment, however, the second priority of Australians was promoting team work and the last priority was the improvement of new methods of assessment (Fig. 6.5).

The third theme of the pedagogical factor reported by participants was designing new assessment methods in the e-learning system. The results of the interviews indicated that there are 4 main requirements in relation to this theme. Both Americans and Australians reported that designing new and novel methods of assignment, assessment with case based project, self-assessment and assessment based on group thread discussion needed to be improved in their e-learning system. Both participants in Australia and America believed that the time and energy that course providers should spend on creating new and novel methods of assessment could be minimized; instead they should put their focus and their effort more on the pedagogical content of the assessment rather than constructing a new method of assessment. Also, the similarity of the requirements reported by participants in both countries reveals that the structure, shape and the program of e-learning (health science) were the same in both countries so the approach to program assessment should be similar in both countries. As has been described in previous themes, both countries had a constructivism perspective about their approach to learning and effective learning practice. Based on these previous themes, it could be stated that in both countries all requirements in relation to assessment were towards a constructivism perspective.

The point that the method of assessment should be more practical than traditional can demonstrate constructivism in both countries. In this regard, "The Island"

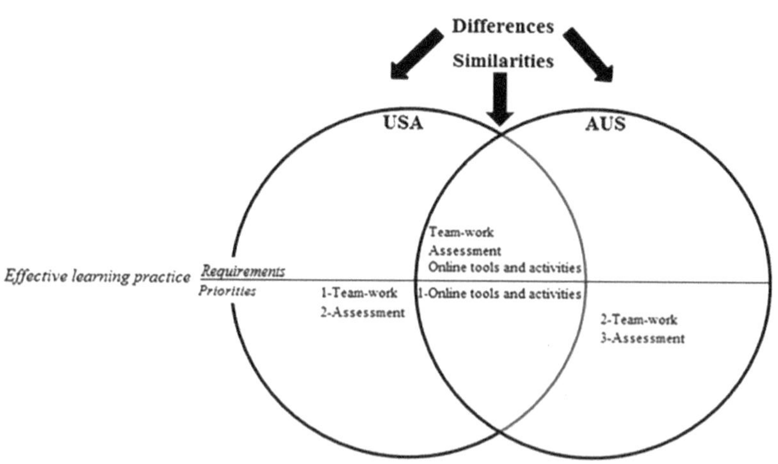

Fig. 6.5 The comparative result based on effective learning practice

assessment was one of the most effective methods of group assessment reported by the American course providers. It is worth suggesting that "The Island" group method of assessment would be applicable to Australian universities as well (Fig. 6.6).

Although Americans believed in applying new methods of group assessment, they also pointed out that supplying these group methods of assessment should be rational and judicious and the point of pedagogy is more important than the construction of new assessment technological bases.

The priority of the need to improve factor is different in America and Australia. From this perspective the findings indicated that in Australia, participants claimed that the first issue that should be considered in the field of assessment is designing new and novel methods of assignment. Australian then claimed that the second issue that needs to improve is group thread discussion. However, in America, people believed that the first requirement in assessment that needs to be improved is group thread discussion and then the second priority is designing new and novel assignments. The third and fourth priorities that are required to be improved were the same in both countries, being case based projects and self-assessment respectively.

The last theme in the pedagogy issue reported by participants was effective learning content. The results of this study showed that there are 3 main requirements namely application of multimedia content, online module content and text content in e-learning systems in both countries. However, the preference of Americans to evaluate multimedia as the most effective learning content was higher than that of Australians. Both course providers in both counties believed that application of multimedia and online modules would be more effective in learning rather than traditional text content. Also both countries believed that the

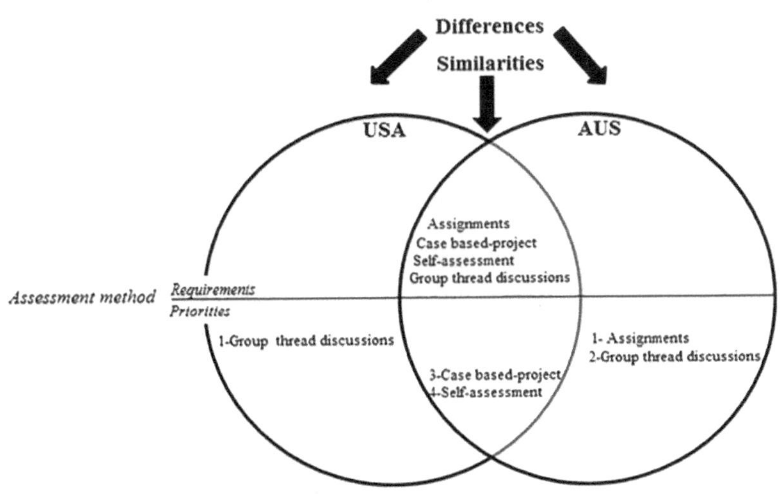

Fig. 6.6 The comparative result based on assessment method

combination of module online content and text content would be more effective than using them separately (Fig. 6.7).

Both countries agreed that the first priority that should be considered by course providers is improvement of multimedia content and the second priority that needs to improve goes to updating online modules and text contents.

The second component arising from the answers of interviewees emphasized cultural issues in e-learning systems in America and Australia. The findings showed that there are 3 main themes in relation to cultural issues. Accordingly, some people believed that cultural issues exist in their e-learning system however, some participants believed that there is non-existence of cultural issues in their system, although the amount of people who believed that the cultural issues existed was higher. The third theme derived from the answers referred to the effective cultural practice that can be used in e-learning system.

Both countries explained the existence of cultural issues in terms of lack of international context awareness and communication issues. This shows that both educational systems, because they have international audiences, have sensitive cultural issues that should be considered in their e-learning system. Considering international students as the recipients of e-learning courses emphasizes that most of them have English as their second language which can make communication difficult.

The findings showed that Americans who believed in the existence of cultural issues also identified the reasons for differences in kinds of attitude as well. This shows that attitudes and perspectives of people in America are also considered as cultural components which should be addressed in e-learning systems; however, Australians didn't mention this which means that the advancement of the e-learning educational system is higher in America than Australia. In Australia, engaging with new technological changes was put forward as the other reason for the existence of

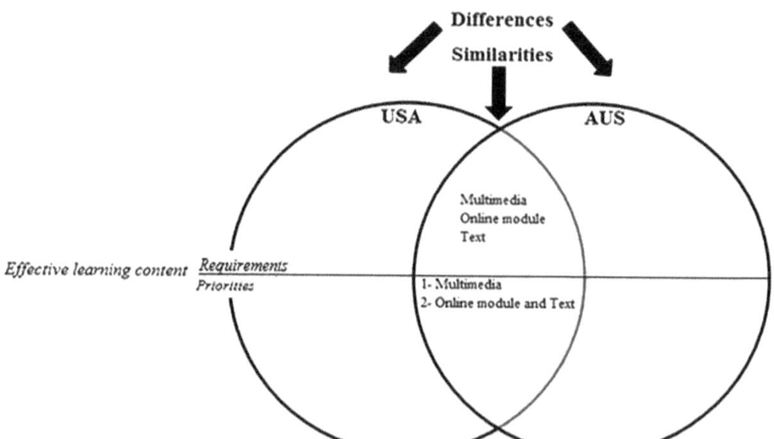

Fig. 6.7 The comparative result based on effective learning content

cultural issues in the e-learning system; not considering this issue in America showed that Americans have passed and solved this issue of the cultural component which again means that Americans have progressed further forward in the management of their e-learning system than Australians. Having challenges in technological issues may lead to a difficult adaptation process for students with a new system. Indeed solving this issue in Australia should be considered by course designers.

In terms of the ranking of these reasons, the Americans explained that the most important issue is communication, the second important issue is lack of awareness and the last important reason regarding the existence of cultural issues belongs to differences in attitude. However, Australians explained that from their perspective the first important reason is lack of awareness, the second reason is communication issues and the last important issue is new technological challenges (Fig. 6.8).

In contrast, the people who do not believe cultural issues exist in their e-learning system gave two reasons for their opinion. The first reason was the similarity in the content and quality of health science courses world wide and the second reason was the global popularity of the courses. These results may be due to the fact that the researcher selected only health science participants in both countries to take part in the study. It can be understandable that the content of health science would be the same in those countries regardless of considering cultural issues (Fig. 6.9).

To continue, the people in both countries explained that the effective cultural practices that are required to be improved are upgrading chat room groups, providing facilities and renovating the e-environment and improving training and skills for both course providers and students. This similarity shows that both countries have the same requirements in terms of effective cultural practice.

Although these effective cultural practice requirements were the same in both countries, the priority of the requirements that needed to be improved was different.

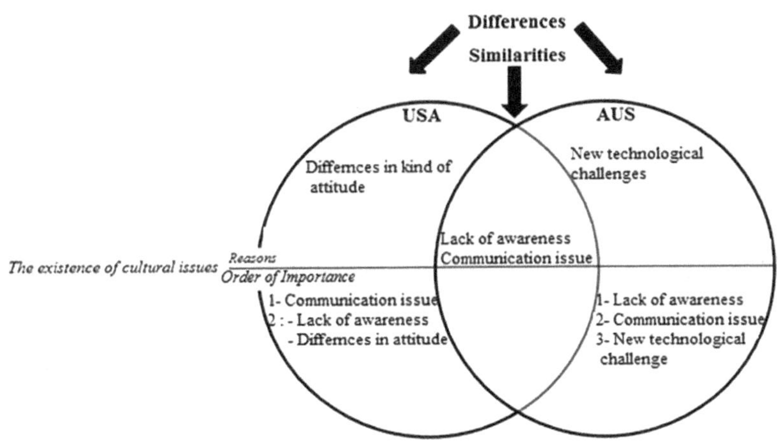

Fig. 6.8 The comparative result based on the existence of cultural issues

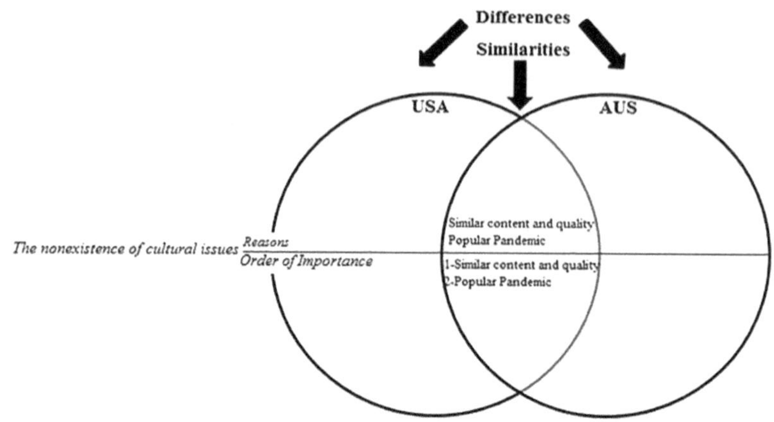

Fig. 6.9 The comparative result based on the nonexistence of cultural issues

In America the first cultural practice that needed to be considered by course providers was improving training and skills of course providers and suppliers. However, this theme was the last priority of Australians. Possibly the fast progress in technology in America was due to the suppliers improving their skills with new training to adapt to technological progress. This means that the progress of technology in Australia in terms of e-learning should be speeded up to reach America.

The second priority of Americans in terms of cultural practice was about updating chat rooms and the last priority belonged to making progress in facilities and renovating the e-environment. However, in Australia, people explained that the second theme that should be improved is making progress in facilities and renovating the e-environment and at the end they placed updating chat rooms (Fig. 6.10).

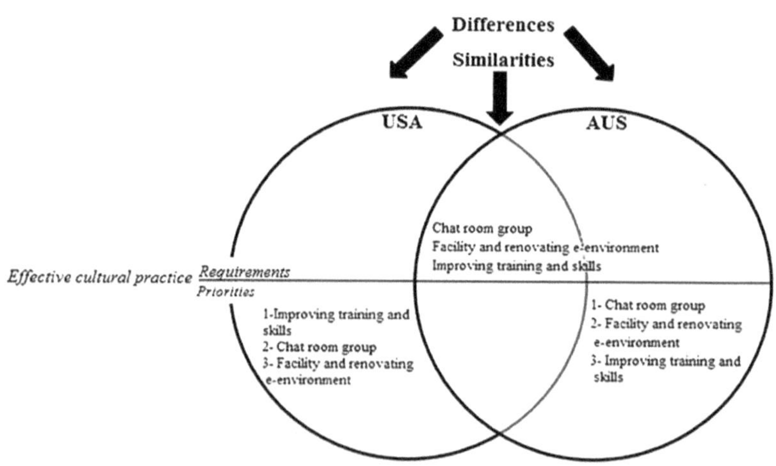

Fig. 6.10 The comparative result based on effective cultural practice

The third factor investigated in the interviews was technological issues. In this part participants evaluated the level of infrastructure of the technology, and the level of functionality of the e-learning system, and then they explained the key technological challenges that existed in their e-learning system. The findings showed that Americans evaluated both infrastructure and functionality of technology higher than Australians. However, Australians believed that the level of these two elements of technology in their e-learning system is low and moderate. These findings are further evidence to show that the experience and history of applying e-learning in America is more progressive than in Australia.

The key technological challenges that have been reported as the same in the two countries were lack of training and development of technological issues, lack of pedagogy driven design and the poor service of the help desk. Although in America lack of updating technologies has been reported as another challenge, in Australia, having old and clunky infrastructures as well as networking issues have been reported as another challenge. These differences in explaining key technological challenges in those countries also revealed that the strong progress of technology in America is greater than in Australia. It seems that old and clunky systems and networking would not be a key challenge in America as it seems to be in Australia.

The first key technological challenge in America that should be considered is the lack of pedagogy driven design then improving the service help-desk. After that comes updating technology of e-learning and at the end making progress in training and skills of suppliers of e-learning.

However, in Australia the first key technological challenge was the need to improve training and technological skills of e-learning suppliers, then updating old or clunky infrastructure, after that addressing the lack of pedagogy driven design and then improving service of help desk and at the end progressing of the network. It seems that the first priorities of Americans are in relation to improving the pedagogical content of the course rather than the technological structure of the course and also poor help desk services. This shows the high level of progress and understanding of course providers in relation to the higher value of pedagogy rather than technology. This understanding however, is lowlighted in Australia as they claim this issue only as their third priority (Fig. 6.11).

The third factor asked about e-learning in America and Australia concerned the best aspect of e-learning in those countries. Both countries defined the best aspect of their e-learning system as flexibility, accessibility and having a self-efficacy environment. In America, providing minimum costs of applying e-learning and in Australia, a collaborative problem solving environment in the e-learning system were mentioned as other best aspects of e-learning.

The ranking of best aspect of e-learning in those countries was firstly flexibility for both countries. Then in America the second best aspect was accessibility and then self-efficacy environment and at the end minimum costs. In Australia, the second best aspect of e-learning was collaborative problem solving then its accessibility and at the end self-efficacy environment (Fig. 6.12).

In the interviews, the fourth factor of interest concerned the areas of e-learning that needed to be improved in overall view. The researcher found that both

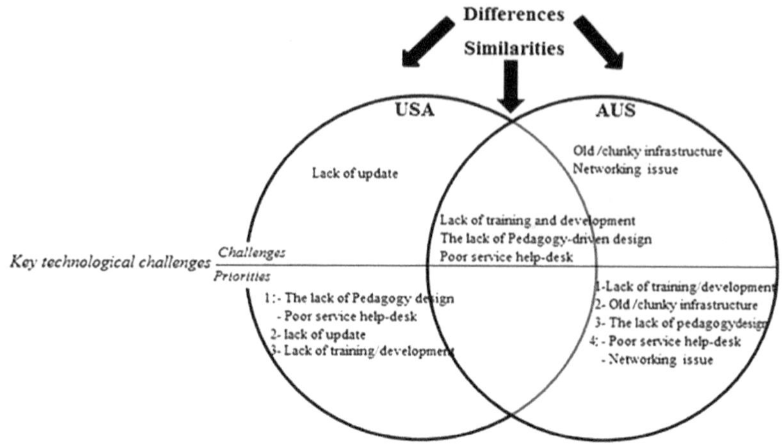

Fig. 6.11 The comparative result based on key technological challenges

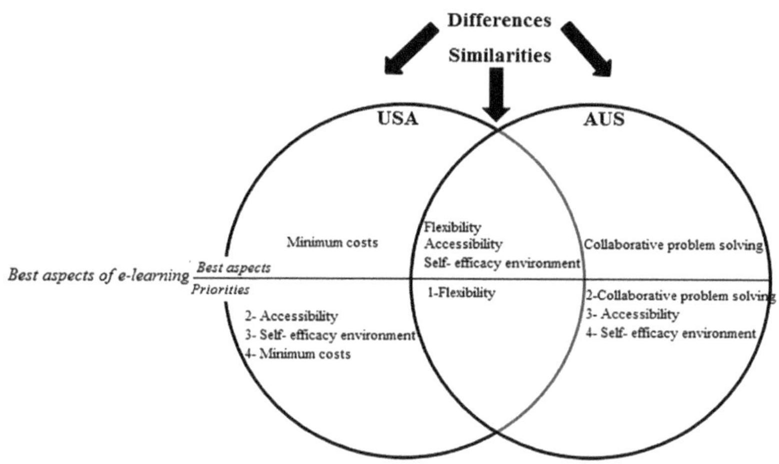

Fig. 6.12 The comparative result based on best aspects of e-learning

countries believed that the policy and procedures of their faculties, the quality of materials that they use, LSM design and its implementation and online support assistance must improve. The empowerment of staff was also mentioned as an issue that just Australians struggled with. Both countries agreed the first thing that needs to improve is policy and procedure and then updating the quality of material. This shows that in the aspect of administrative support practice (need to improve policy and procedure, and, just in Australia, empowerment of staff) and in the aspect of pedagogical practice (need to update the quality of materials) and in the aspect of technological practice (updating LSM design and its implementation, online support assistance) both countries need to make progress. Indeed both content and structure of e-learning systems should be updated (Fig. 6.13).

The first priority considered in comments from both countries was faculty policies and procedure and the second priority was improving the quality of materials used. This indicates that the first priority should focus on administrative support practices and then pedagogical practices which were the same in both countries. The last priority however, concerned improving technological practice. In this case, in America, the third issue that needed to improve was LSM design and online support assistance and in Australia as second priority LSM design and its implementation and thirdly online support system as well as empowerment of staff.

In general as can be seen in Table 6.10, the results of this study have focused on four main practices including pedagogy, culture, technology and aspects that need to be improved. In terms of pedagogy practice this research found that approaches to learn such as collaborative learning, problem solving and feedback base are similar requirements in both Australia and America. However, in America the outcome base approach to learning was also mentioned as an important requirement in e-learning systems. In addition, requirements such as team working, assessment, and online tools have been mentioned by both countries as effective learning practices. Both countries also argued that assessment methods like assignment, case based projects, self-assessment, and discussion groups need to be developed in e-learning systems. Also both countries believed that learning contents like multimedia, online module, and texts in e-learning systems need to be updated. These findings revealed that the pedagogical issues and concerns in both countries are the same which shows that an e-learning system regardless of different countries needs more improvement in pedagogy.

In terms of culture practice, cultural sensitivity like lack of awareness and communication issues should be considered in both countries. Concerning this, different kinds of attitudes in America and new technological challenges in

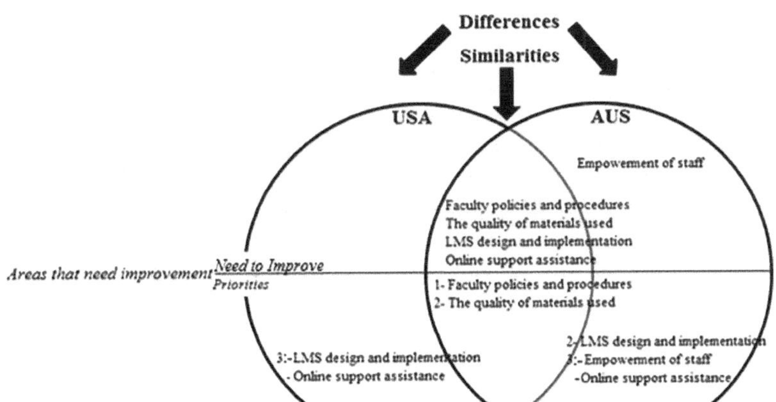

Fig. 6.13 The comparative result based on area that need improvement

Table 6.10 The comparative results of one faculty in an Australian University and one faculty in a US University

Main practices	Current issues	Similarities	USA participants	AUS participants
Pedagogy	Approaches to learning	Collaborative, problem solving and feedback based approaches required	Outcome based approaches required	–
	Effective learning practice	Team work, Assessment and online tools required	–	–
	Assessment method	Assignment, case based project, self-assessment and discussion group required	–	–
	Learning content	Multimedia, online module and text required to updates	–	–
Culture	Cultural sensitivity	Lack of awareness and communication issues	Differences in kind of attitude	New technological challenges
	Effective cultural practice	Chat room systems, new online facilities and improving training required	–	–
Technology	Technological challenges	Lack of training, Lack of pedagogical design, poor service help desk	Lack of update	Old systems and networking issues
Needs to be improve	Areas which need to be improved in e-practice system	Faculty polices, Quality, LMS, Online support	–	Empowerment of staff

Australia also have been mentioned. For effective cultural practices, both countries explained that updating chat room systems, new online facilities and improving training are required.

In terms of technology challenges, this research found that technological challenges like lack of training, lack of pedagogical design, poor service help desk are the same in both countries and need to improve. Also lack of updating the e-learning system in America and old systems and networking issues in Australia have been described as the technological challenges which need to improve. The findings showed that the level of technology in terms of infrastructure and functionality of e-learning is higher in America than Australia. As the interviews with Australians explained, they are dissatisfied with the aspect of technology in e-learning which needs to be progressed and updated.

Finally, this research summarized all areas that needed to improve in e-practice systems. Both countries considered that faculty policies and procedures, quality of materials, LMS, and online support assistance need to improve. In addition, empowerment of staff was mentioned by Australians.

6.4.1 Limitations

There are some limitations in regard to the conducting of this study. Interviews were the only method used to collect the data. The limited time frame and financial budget to conduct the study as well as the distance between Australia and America did not allow the researcher to have face to face meetings with American participants. As mentioned, all data collected from America, were based on using telephone and Skype. Future research should focus on investigating current e-learning practice in America and Australia using observation and focus group methodology.

Another limitation concerns accessibility to document analysis. Current issues of e-learning practice would have provided useful data. However, current documents with regard to e-learning practice were under the process of being written at the time that the research was conducted. As a result they were unavailable. Further, due to confidentiality of documents, universities were not inclined to give them to the researcher to report on.

Another limitation was the difficulty of access to administrative staff. Most of them were very busy and didn't have any interest in taking part in this study.

In general this study attempted to investigate in a deeper perspective some aspects of study 1 in Chap. 3 (cultural context of e-learning practice based on learning preferences) and study 2 in Chap. 4 (assessment of e-practice factors in one faculty in an Australian university and one faculty in a US university). Ideally, the outcome of this research would be the creation of a collaborative committee between Australia and America to work jointly on shared aspects of e-learning challenges to solve them and strengthen the best aspects of e-learning. This committee would help both countries exchange successful experiences and apply them in their different cultural contexts. Support from the technological expertise of the American university sample would help the Australian university sample to make progress in this area.

6.5 General Discussion

The main aim of this research study was to provide a comparative understanding of e-practice factors in an Australian and an American university. In the first study the dominant cultural dimensions between one faculty in an Australian university and one faculty in a US university were discussed which showed that the dominant cultural aspect of participants of one faculty in an American university was toward constructivism while the dominant cultural aspect of participants of one faculty in an Australian university was toward instructivism. The main explanation for the results may refer to the fact that most suppliers of the Australian university e-learning system (including administrative and academic staff of universities) were from eastern Asian cultural background with a dominant traditional perspective about learning. In Study 2, findings indicated that the levels of e-practice factors in one faculty in the Australian university and one faculty in the US university were above average. However, in the American university the terms of e-practice were evaluated more highly. In Study 3, the current issues and problems of e-learning practice in 4 aspects of pedagogy, culture, technology and e-practice area which need to improve have been investigated and it was found that cultural sensitivity and effective cultural practice, key technological challenges and issues like faculty policies, quality, LMS, and online support need to improve. In general this research study suggested that it is essential for the Australian university sample to develop, update and sustain the e-learning educational system especially in terms of e-practice using the scientific pattern and technology that other developed countries like America as pioneers are applying. This updating and sustaining may be more necessary for the Australian university sample (see Isaias and Issa 2013; Chigeza and Halbert 2014), although it is also important to remember that copying those patterns without considering the Australian culture and the audiences that are using the e-learning system may negatively impact the effectiveness of learning. Indeed the combination of adopting successful patterns of other developed countries as well as adjusting them to fit with the Australian culture would be the best strategy for educational decision and policy makers of the future.

References

Achtemeier, S. D., & Simpson, R. D. (2005). Practical considerations when using benchmarking for accountability in higher education. *Innovative Higher Education, 30*(2), 117–128.

Academic Affairs & Provost. (2016, August 01). Retrieved from http://www.academic.umn.edu/provost/elearning/.

Adamson, B. (2012). International comparative studies in teaching and teacher education. *Teaching and teacher education, 28*(5), 641–648.

Adelsberger, H. H., Collis, B., & Pawlowski, J. M. (Eds.). (2013). *Handbook on information technologies for education and training*. Springer Science & Business Media.

Aguti, B., Walters, R., & Wills, G. (2014). Effective use of e-learning technologies to promote student-centered learning paradigms within higher education institutions. *International Journal for e-Learning Security (IJeLS)*, 1–8. Retrieved from http://eprints.soton.ac.uk/382471/1/Effective%20Use%20of%20E%20Learning%20Technologies%20to%20Promote%20Student%20Centered%20Learning.pdf.

Alexander, R., Broadfoot, P., & Phillips, D. (1999). *Learning from comparing: New directions in comparative education research.* Symposium Books.

Algers, A., Lindström, B., & Pajor, E. A. (2011). A new format for learning about farm animal welfare. *Journal of Agricultural and Environmental Ethics, 24*(4), 367–379. Retrieved from http://pub.epsilon.slu.se/11219/1/algers_et_al_140526.pdf.

Al-Huneidi, A., & Schreurs, J. (2011). Constructivism based blended learning in higher education. In *World Summit on Knowledge Society*(pp. 581–591). Springer Berlin Heidelberg. Retrieved from https://doclib.uhasselt.be/dspace/bitstream/1942/12273/1/Constructivism%20Based%20Blended%20Learning%20in%20Higher%20Education-AL-HUNIEIDI%20Ahmad-Hasselt%20University.pdf.

Ali, N. S., Hodson-Carlton, K., & Ryan, M. (2004). Students' perceptions of online learning: Implications for teaching. *Nurse Educator, 29*(3), 111–115.

Allen, I. E., & Seaman, J. (2007). *Making the grade: Online education in the United States, 2006.* Sloan Consortium. PO Box 1238, Newburyport, MA 01950.

Allen, I. E., & Seaman, J. (2011). *Going the distance: Online education in the United States, 2011.* Sloan Consortium. PO Box 1238, Newburyport, MA 01950. Retrieved from http://files.eric.ed.gov/fulltext/ED529948.pdf.

Allen, I. E., & Seaman, J. (2015). Grade level: Tracking online education in the United States. *Babson Park, MA: Babson Survey Research Group. Accessed March, 10,* 2015. Retrieved from http://www.onlinelearningsurvey.com/reports/gradelevel.pdf.

Alley, L. R., & Jansak, K. E. (2001). The ten keys to quality assurance and assessment in online learning. *Journal of Interactive Instruction Development, 13*(3), 3–18.

S.H. Sadeghi, *E-Learning Practice in Higher Education: A Mixed-Method Comparative Analysis*, Studies in Systems, Decision and Control 122,
https://doi.org/10.1007/978-3-319-65939-8

Alzahrani, J. G., & Ghinea, G. (2012, June). Evaluating the impact of interactivity issues on e-learning effectiveness. In *2012 International Conference on Information Technology Based Higher Education and Training (ITHET)* (pp. 1–5). IEEE. Retrieved from http://repository.taibahu.edu.sa/bitstream/handle/123456789/5041/Evaluating%20the%20impact%20of%20interactivity%20issues%20on%20e-learning%20effectiveness.pdf?sequence=1.

Andersson, A. S., & Grönlund, Å. (2009). A conceptual framework for e-learning in developing countries: A critical review of research challenges. *The Electronic Journal of Information Systems in Developing Countries, 38*. Retrieved from file:///C:/Users/ssad2473/Downloads/564-1350-1-PB.pdf.

Anjorin, M., Rodenhausen, T., García, R. D., & Rensing, C. (2012, September). Exploiting semantic information for graph-based recommendations of learning resources. In *European Conference on Technology Enhanced Learning* (pp. 9–22). Berlin Heidelberg: Springer. Retrieved from ftp://www.kom.tu-darmstadt.de/papers/ARDR12-1.pdf.

Antoine, J. E. (2011). *E-Learning: A Student's Perspective a Phenomenological Investigation*. ProQuest LLC. 789 East Eisenhower Parkway, PO Box 1346, Ann Arbor, MI 48106.

Arnove, R. F., Torres, C. A., & Franz, S. (Eds.). (2012). *Comparative education: The dialectic of the global and the local*. Rowman & Littlefield Publishers.

Ary, D., Jacobs, L. C., Razavieh, A., & Sorenson, C. (2006). *Introduction to Research in Education*. Belmont: CA, Thomson Wadsworth.

Bagnall, N. (2015). *Global Identity in Multicultural and International Educational Contexts: Student Identity Formation in International Schools*. Routledge.

Barber, J. (2013). E-learning: Supplementary or disruptive?.*Telecommunications Journal of Australia, 63*(1).

Barker, K. (1999). Quality guidelines for technology-assisted distance education. *Washington, DC: US Department of Education Office of Learning Technologies*. Retrieved from http://tecfa.unige.ch/proj/cvs/doc_ress/FutureEdGuide1.pdf.

Bates, T. (1997). *Restructuring the university for technological change* (pp. 78–101). Murdoch University.

Bednar, A. K. (1992). *Attempting to come to grips with alternative perspectives* (p. 129). Constructivism and the technology of instruction: A conversation.

Bennett, S., Maton, K., & Kervin, L. (2008). The 'digital natives' debate: A critical review of the evidence. *British Journal of Educational Technology, 39*(5), 775–786. Retrieved from http://ro.uow.edu.au/cgi/viewcontent.cgi?article=2465&context=edupapers.

Bereday, G. Z. (1964). *Comparative Method in Education, Holt*. USA: Rinehart and Winston. Inc.

Bignold, W., & Gayton, L. (Eds.). (2009). *Global issues and comparative education*. Learning Matters.

Bjekic, D., Krneta, R., & Milosevic, D. (2010). Teacher education from e-learner to e-teacher: Master curriculum. *TOJET: The Turkish Online Journal of Educational Technology, 9*(1). Retrieved from http://files.eric.ed.gov/fulltext/EJ875783.pdf.

Boud, D., & Molloy, E. (2013). *Feedback in higher and professional education: understanding it and doing it well*. Routledge.

Bowles, M. S., & Bowles, M. S. (2004). *Relearning to e-learn: Strategies for electronic learning and knowledge*. Academic Monographs.

Bramble, W. J., & Panda, S. (2008). Organizational and cost structures for distance and online learning. *Economics of distance and online learning. Theory, Practice, and Research*, 1–12.

Bray, M., Adamson, B., & Mason, M. (Eds.). (2014). *Comparative education research: Approaches and methods* (Vol. 19). Springer.

Breitner, M. H., & Hoppe, G. (2005). A glimpse at business models and evaluation approaches for e-learning. *E-learning*, 179–193. Retrieved from http://citeseerx.ist.psu.edu/viewdoc/download?doi=10.1.1.198.5553&rep=rep1&type=pdf.

Buckland, M., & Dye, C. M. (1991). The Development of Electronic Distance Education Delivery Systems in the United States. *Recurring and Emerging Themes in History and Philosophy of Education*. Retrieved from http://eric.ed.gov/?id=ED345713

Butcher, N., & Wilson-Strydom, M. (2013). A guide to quality in online learning. Dallas, TX: Academic Partnerships.

Butter, M. C., Valenzuela, E. S., & Quintana, M. G. B. (2015). Intercultural Talent Management Model: Virtual communities to promote collaborative learning in indigenous contexts. Teachers' and students' perceptions. *Computers in Human Behavior, 51,* 1191–1197. doi:10.1016/j.chb.2015.01.030.

Chaney, B. H. (2004). *History, theory, and quality indicators of distance education: A literature review*. Office of Health Informatics: Texas A&M University.

Chen, M. L., Su, Z. Y., Wu, T. Y., Shieh, T. Y., & Chiang, C. H. (2011). Influence of dentistry students'e-Learning satisfaction: a questionnaire survey. *Journal of medical systems, 35*(6), 1595–1603.

Chen, S. J., Caropreso, E. J., Hsu, C. L., & Yang, J. (2012). Cross-cultural collaborative online learning: If you build it, will they come? *Global Partners in Education Journal*, *2*(1), 25. Retrieved from http://www.gpejournal.org/index.php/GPEJ/article/view/46

Chickering, A. W., & Ehrmann, S. C. (1996). Implementing the seven principles: Technology as lever. *AAHE Bulletin*, *49*, 3–6. Retrieved from http://www.fmtsystems.com/04-news/Impl-7-prin.pdf.

Chickering, A. W., & Gamson, Z. F. (1987). Seven principles for good practice in undergraduate education. *AAHE bulletin, 3,* 7.

Chigeza, P., & Halbert, K. (2014). Navigating E-Learning and Blended Learning for Pre-Service Teachers: Redesigning for Engagement, Access and Efficiency. *Australian Journal of Teacher Education*, *39*(11), n11. Retrieved from http://files.eric.ed.gov/fulltext/EJ1047088.pdf.

Chou, C. (2003). Interactivity and interactive functions in web-based learning systems: a technical framework for designers. *British Journal of Educational Technology*, *34*(3), 265–279. Retrieved from https://ir.nctu.edu.tw/bitstream/11536/14462/1/000183171200004.pdf.

Chou, S. W., & Liu, C. H. (2005). Learning effectiveness in a Web-based virtual learning environment: a learner control perspective. *Journal of Computer Assisted Learning*, *21*(1), 65–76. Retrieved from https://www.researchgate.net/profile/ShihWei_Chou/publication/220663188_Learning_effectiveness_in_a_Webbased_virtual_learning_environment_a_learner_control_perspective/links/54eb2d9b0cf27a6de1174fbc.pdf.

Coiro, J. (2003). Exploring literacy on the internet: Reading comprehension on the internet: Expanding our understanding of reading comprehension to encompass new literacies. *The Reading Teacher*, *56*(5), 458–464. Retrieved from http://stancock.iweb.bsu.edu/edrdg445/online/pdf/coiro.pdf.

Collis, B. (1999). Designing for differences: Cultural issues in the design of WWW-based course-support sites. *British Journal of Educational Technology, 30*(3), 201–215. doi:10.1111/1467-8535.00110.

Cottrell, S., & Donaldson, J. H. (2013). Exploring the opinions of registered nurses working in a clinical transfusion environment on the contribution of e-learning to personal learning and clinical practice: results of a small scale educational research study. *Nurse Education in Practice, 13*(3), 221–227.

Commissions, R. A. (2001). Best practices for electronically offered degree and certificate programs. Supplements Guidelines for Distance Education. Retrieved from http://www.niu.edu/assessment/manual/_docs/Best%20Practices.pdf.

Creswell, J. W. (2013). *Qualitative inquiry and research design: Choosing among five approaches*. Sage.

Creswell, J. W., & Clark, V. L. P. (2007). Designing and conducting mixed methods research. Sage.

De Freitas, S., & Oliver, M. (2005). Does E-learning Policy Drive Change in Higher Education?: A case study relating models of organisational change to e-learning implementation. *Journal of Higher Education Policy and Management, 27*(1), 81–96. doi:10.1080/13600800500046255.

Denscombe, M. (2014). *The good research guide: For small-scale social research projects*. McGraw-Hill Education (UK).

Devlin, M . (2015) *Learning and teaching plan 2015-2017* . Enabling BOLD learning . Federation University Australia. Retrieved from http://federation.edu.au/__data/assets/pdf_file/0007/227581/Learning-and-Teaching-Plan-2015-17_FINAL.pdf.

Díez, D., Díaz, P., & Aedo, I. (2012). The ComBLA method: The application of domain analysis to the development of e-learning systems. *Journal of Research and Practice in Information Technology, 44*(3), 331.

DIISRTE (2013) *Higher education data cube (uCube).* Higher education statistics. Retrieved from http://www.highereducationstatistics.deewr.gov.au/Default.aspx.

Dillenbourg, P. (1999). What do you mean by collaborative learning. *Collaborative-Learning: Cognitive and Computational Approaches, 1*, 1–15.

Dirckinck-Holmfield, L., Jones, C., & Lindström, B. (2009). Analysing networked learning practices in higher education and continuing professional development. *Sense Publishers.*

Dirr, P. J. (2003). Distance education policy issues: Towards 2010. *Handbook of distance education* (pp. 461–479). Retrieved from http://www.contaduria.uady.mx/sec-cip/articulos/libros_online/educacion/LawrenceErlbaum2003HandbookOfDistanceEducation.pdf#page=487.

Doolittle, P. (1999). Constructivism and online education.

Downing, J., Pittaway, S& Osborne, P. (2014). *Online facilitation and the role of the tutor.* Faculty of Education, University of Tasmania. Retrieved from http://www.utas.edu.au/__data/assets/pdf_file/0005/202748/Online-facilitation-and-the-role-of-the-Tutor.pdf.

Dragon, T., Mavrikis, M., McLaren, B. M., Harrer, A., Kynigos, C., Wegerif, R., et al. (2013). Metafora: A web-based platform for learning to learn together in science and mathematics. *IEEE Transactions on Learning Technologies, 6*(3), 197–207.

Duffy, T. M., & Jonassen, D. H. (Eds.). (2013). *Constructivism and the technology of instruction: A conversation*. Routledge.

Duffy, T. M., Lowyck, J., & Jonassen, D. H. (2012). *Designing environments for constructive learning*. Springer.

Edmundson, A. (2003). Decreasing cultural disparity in educational ICTs: Tools and recommendations. *Turkish Online Journal of Distance Education, 4*(3). Retrieved from http://dergipark.ulakbim.gov.tr/tojde/article/view/5000102991/0.

Edmundson, A. (2004). *The cross-cultural dimensions of globalized e-learning*. PhD Thesis, Walden University, USA.

Ehlers, U. D. (2009). Understanding quality culture. *Quality Assurance in Education, 17*(4), 343–363. doi:http://dx.Doi.org/10.1108/09684880910992322

Elias, M. (1997). The missing piece. *Education Week, 17*(15), 36–38.

Ellis, R. A., Ginns, P., & Piggott, L. (2009). E-learning in higher education: some key aspects and their relationship to approaches to study. *Higher Education Research & Development, 28*(3), 303–318.Retrieved from http://www.informaworld.com/10.1080/07294360902839909.

Fichten, C. S., Asuncion, J. V., Barile, M., Ferraro, V., & Wolforth, J. (2009). Accessibility of e-learning and computer and information technologies for students with visual impairments in postsecondary education. *Journal of Visual Impairment and Blindness, 103,* 543–557.

Finger, G., Jamieson-Proctor, R., & Watson, G. (2006). Measuring learning with ICTs: an external evaluation of Education Queensland's ICT curriculum integration performance measurement instrument. In *Proceedings of the Australian Association for Research in Education Annual Conference (AARE 2005)*. Australian Association for Research in Education.

Fischer, H., Heise, L., Heinz, M., Moebius, K., & Koehler, T. (2015). How to identify e-learning trends in academic teaching: Methodological approaches and the analysis of scientific discourses. *Interactive Technology and Smart Education, 12*(1), 31–43.

FitzPatrick, T. (2012). *Key Success Factors of eLearning in Education: A Professional Development Model to Evaluate and Support eLearning*. Online Submission.

Flyvbjerg, B. (2006). Five misunderstandings about case-study research. *Qualitative Inquiry, 12* (2), 219–245. Retrieved from http://qix.sagepub.com/content/12/2/219.

Forsyth, H., Pizzica, J., Laxton, R., & Mahony, M. J. (2010). Distance education in an era of eLearning: challenges and opportunities for a campus-focused institution. *Higher Education Research & Development, 29*(1), 15–28.

Fraser, B. J., & Fisher, D. L. (1994). Assessing and researching the classroom environment. *The study of learning environments* (Vol. 8, pp. 23–38).

Fry, K. (2001). E-learning markets and providers: some issues and prospects. *Education+ Training, 43*(4/5), 233–239. Retrieved from http://www.emeraldinsight.com/doi/abs/10.1108/EUM0000000005484.

Frydenberg, J. (2002). Quality standards in eLearning: A matrix of analysis. *The International Review of Research in Open and Distributed Learning, 3*(2). Retrieved from http://www.irrodl.org/index.php/irrodl/article/viewArticle/109

Frymier, A. B. (2005). Students' classroom communication effectiveness. *Communication Quarterly, 53*(2), 197–212.

Gagne, R. M., Wager, W. W., Goals, K. C., & Kelle, J. M. (2005). Principle of instructional design. Belmont, CA: Thomson Learning. *Educational Multimedia, Hypermedia & Telecommunications, 2002,* 1251–1257.

Gamble, A. L. (2009). *The effectiveness of e-learning in a multicultural environment.* PhD Theses. Capella University United States – Minnesota. Retrieved from http://search.proquest.com/docview/288109233.

Garrison, D. R. (2011). *E-learning in the 21st century: A framework for research and practice.* Taylor & Francis.

Gáti, J., & Kártyás, G. (2011, September). Practice oriented higher education course definitions and processes. In *2011 5th International Symposium on Computational Intelligence and Intelligent Informatics (ISCIII)* (pp. 127–131). IEEE.

Gerjets, P. H., & Hesse, F. W. (2004). When are powerful learning environments effective? The role of learner activities and of students' conceptions of educational technology. *International Journal of Educational Research, 41*(6), 445–465.

Glaser, B. G. (1992). *Emergence vs. forcing. advances in the methodology of grounded theory* . Mill Valley, CA: Sociology Press.

Glaser, B., & Strauss, A. L. (1967). *The discovery of grounded research: strategies for qualitative research.* New Brunswick (USA) and London (UK): Aldine Transaction.

Gokhale, A. A. (1995). Collaborative learning enhances critical thinking. *Journal of Technology Education, 7*(1), 5.

González-Videgaray, M. (2007). Evaluación de la reacción de alumnos y docentes en un modelo mixto de aprendizaje para educación superior. *RELIEVE, 13*(1) Retrieved fromhttp://www.uv.es/RELIEVE/v13n1/RELIEVEv13n1_4.htm.

Grant, N. (1999). *Comparing educational systems. An Introduction to education studies.* London: David Fulton.

Gulati, S. (2008). Technology-enhanced learning in developing nations: A review. *The International Review of Research in Open and Distributed Learning, 9*(1). Retrieved from http://www.irrodl.org/index.php/irrodl/article/viewArticle/477/1012.

Gunawardena, C. N., Wilson, P. L., & Nolla, A. C. (2003). Culture and online education. *Handbook of distance education* (pp. 753–775). Retrieved from http://ltc-ead.nutes.ufrj.br/constructore/objetos/Handbook_Of_Distance_Education.pdf#page=778.

Guo, Y., Qian, D., Guan, J., & Wang, J. (2010). Usability testing on a government training platform: A case study. In *2010 2nd International Conference on Education Technology and Computer* (Vol. 2, p. V2-211). IEEE.

Guo, Y., Wang, J., Moore, J., Liu, M., & Chen, H.-L. (2009). A case study of usability testing on an asynchronous e-Learning platform. Paper presented at the *2009 Joint Conferences on Pervasive Computing (JCPC).*

Guo, Y., Wang, J., Moore, J., Liu, M., & Chen, H. L. (2009). A case study of usability testing on an asynchronous e-learning platform. In *2009 Joint Conferences on Pervasive Computing (JCPC)* (pp. 693–698). IEEE.

Guribye, F. (2005). Infrastructures for learning. *Ethnographic inquiries into the social and technical conditions of education and training*. The University of Bergen.

Guribye, F. (2015). From artifacts to infrastructures in studies of learning practices. *Mind, Culture, and Activity, 22*(2), 184–198. doi:10.1080/10749039.2015.1021358.

Hacker, D. J., & Niederhauser, D. S. (2000). Promoting deep and durable learning in the online classroom. *New Directions for Teaching and Learning, 2000*(84), 53–63. doi:10.1002/tl.848.

Hancock, A. (1999). The evolving terrain of distance learning. *Satellite Communications, 23*(3), 1–4.

Hase, S., & Ellis, A. (2001). Problems with online learning are systemic, not technical. In J. Stephenson (Ed.), *Teaching and learning online: pedgogies for new technologies*. London: Kogan Page.

Heafner, T., Hartshorne, R., & Petty, T. (2015). Exploring Online Education in K-12 Environments. In *Society for Information Technology & Teacher Education International Conference* (Vol. 1, pp. 1478–1480). Retrieved from http://www.editlib.org/p/150585.

Hein, G. (1991). Constructivist learning theory. *Institute for Inquiry*. Retrieved from http://www.exploratorium.edu/ifi-archive/resources/research/constructivistlearning.html.

Henderson, L. (1996). Instructional design of interactive multimedia: A cultural critique. *Educational Technology Research and Development, 44*(4), 85–104. doi:10.1007/BF02299823.

Henry, M. M. (1973). Methodology in comparative education: An annotated bibliography. *Comparative Education Review, 17*(2), 231–244.

Hensley, D., & Goldsmith, C. (2013). *Good e-practice guidelines for disadvantaged learners in VET*. Flexible Learning Advisory Group. Retrieved from http://flexiblelearning.net.au/about/research/good-e-practice-guidelines-for-disadvantaged-learners-in-vet-final-report-2/.

Herman, N. (2001). *The applicability of international benchmarks to an Internetbased distance education programme*. Thesis: University of Stellenbosch, South Africa.

Herrington, A., Herrington, J., Oliver, R., Stoney, S., & Willis, J. (2001). Quality guidelines for online courses: The development of an instrument to audit online units. In *Proceedings of 18th Conference of the Australasian Society for Computers in Learning in Tertiary Education* (pp. 263–270). Melbourne, VIC. Biomedical Multimedia.

Herrington, J., Oliver, R., & Herrington, A. (2007). Authentic Learning on the Web: Guidelines. *Flexible Learning in an Information Society*, 26–35.

Herriot, A. M., Bishop, J. A., Kelly, M., Murphy, M., & Truby, H. (2003). Evaluation of a computer assisted instruction resource in nursing education. *Nurse Education Today, 23*(7), 537–545.

Hiemstra, R. (2009). Writing objectives, executive summaries, criterion referenced testing, goals for the first team meeting, and stages of team growth. IACE Hall of Fame Repository.

Hills, H., & Overton, L. (2010). *Why do e-learning projects fail? 33 causes of failure (& what to do about them!)*. Towards Maturity CIC 2010. Retrieved from https://www.aoc.co.uk/sites/default/files/Why_Elearning_Projects_Fail.pdf.

Ho, A. D., Reich, J., Nesterko, S. O., Seaton, D. T., Mullaney, T., Waldo, J., et al. (2014). HarvardX and MITx: The first year of open online courses, fall 2012-summer 2013. Ho, AD, Reich, J., Nesterko, S., Seaton, DT, Mullaney, T., Waldo, J., et al. (2014). *HarvardX and MITx: The first year of open online courses (HarvardX and MITx Working Paper No. 1)*.

Hofstede, G. H., & Hofstede, G. (2001). *Culture's consequences: Comparing values, behaviors, institutions and organizations across nations*. Sage.

Holly, C., Legg, T., Mueller, D., & Adelman, D. S. (2008). Online teaching: Challenges for a new faculty role. *Journal of Professional Nursing, 24*(4), 254–258. Retrieved from http://dx.Doi.org/10.1016/j.profnurs.2007.07.003

Holmberg, B., Hrsg. Bernath, & Busch, F. W. (2005). *The evolution, principles and practices of distance education* (Vol. 11). Bis. Retrieved from http://www.uni-oldenburg.de/fileadmin/user_upload/c3l/master/mde/download/asfvolume11_eBook.pdf.

Holsapple, C. W., & Lee-Post, A. (2006). Defining, Assessing, and promoting e-learning success: an information systems perspective. *Decision Sciences Journal of Innovative Education, 4*(1), 67–85.

Hosie, P., Schibeci, R., & Backhaus, A. (2005). A framework and checklists for evaluating online learning in higher education. *Assessment & Evaluation in Higher Education, 30*(5), 539–553.

Hostager, T. J. (2014). Online learning resources do make a difference: mediating effects of resource utilization on course grades. *Journal of Education for Business, 89*(6), 324–332. Retrieved from http://www.tandfonline.com/Doi/abs/10.1080/02602930500187097.

Huang, H. M. (2002). Toward constructivism for adult learners in online learning environments. *British Journal of Educational Technology, 33*(1), 27–37. Retrieved from https://brainmass.com/file/1433703/Article.pdf.

IBS World. (2014). *Online education in Australia*: Market research report. In a. X008 (Ed.), Retrieved from http://www.ibisworld.com.au/industry/default.aspx?indid=1907.

Im, J. H. (2006). Development of an e-education framework. *Online Journal of Distance Learning Administration, 9*(4), 1–9.

Inglis, A. (2005). Quality improvement, quality assurance, and benchmarking: comparing two frameworks for managing quality processes in open and distance learning. *The International Review of Research in Open and Distance Learning, 6*(1). Retrieved from http://www.irrodl.org/index.php/irrodl/article/view/221/304.

Inglis, A. (2008). Approaches to the validation of quality frameworks for e-learning. *Quality Assurance in Education, 16*(4), 347–362. http://dx.Doi.org/10.1108/09684880810906490.

Isaias, P., & Issa, T. (2013). E-learning and sustainability in higher education: an international case study. *International Journal of Teaching and Learning in Higher Education, 20*(4), 77–90. Retrieved from http://find.curtin.edu.au/staff/index.cfm.

Jacobson, M. J., & Spiro, R. J. (1994). A framework for the contextual analysis of technology-based learning environments. *Journal of Computing in Higher Education, 5*(2), 3–32.

Jacobson, M. J., Kim, Y., Lee, J., Kim, H., & Kwon, S. (2005). Learning sciences principles for advanced e-learning systems: implications for computer-assisted language learning. *Multimedia Assisted Language Learning, 8*(1), 76–115.

Jereb, E., & Šmitek, B. (2006). Applying multimedia instruction in e-learning. *Innovations in Education & Teaching International, 43*(1), 15–27.

Jonassen, D. H. (1991). Objectivism versus constructivism: Do we need a new philosophical paradigm? *Educational Technology Research and Development, 39*(3), 5–14.

Juwah, C. (2013). *Interactions in online education: Implications for theory and practice*: Routledge.

Kala, S., Isaramalai, S.-A., & Pohthong, A. (2010). Electronic learning and constructivism: A model for nursing education. *Nurse Education Today, 30*(1), 61–66.

Karabenick, S. A., & Newman, R. S. (Eds.). (2013). *Help seeking in academic settings: Goals, groups, and contexts*. Routledge.

Kvale, S. (1996). *InterViews: An introduction to qualitative research interviewing*. Thousand Oaks, CA: Sage.

Keegan, D. (1990). *A theory for distance education. Contemporary issues in American distance education* (pp. 327–332). Oxford: Pergamon Press,.

Keeton, M. T. (2004). Best online instructional practices: Report of phase I of an ongoing study. *Journal of Asynchronous Learning Networks, 8*(2), 75–100. Retrieved from https://wcpss.pbworks.com/f/BP+Online.pdf.

Kelsey, K. (2007). Constructivism: strength and weakness. Retrieved from http://adulteducation.wikibook.us/index.php?title=Constructivism.

Khan, B. H. (2005). *E-learning quick checklist.* Information Science Pub.

Khan, B. H., & Granato, L. A. (2008). Program evaluation in e-learning: Retrieved from http://asianvu.com/bk/elearning_evaluation_article.pdf.

Klašnja-Milićević, A., Vesin, B., Ivanović, M., & Budimac, Z. (2011). E-Learning personalization based on hybrid recommendation strategy and learning style identification. *Computers & Education, 56*(3), 885–899.

Kolb, D. (1984). *Experiential learning*. Englewood Cliffs, NJ: Prentice-Hall.

Krause, K. L. D., McEwan, C., & Blinco, K. (2009). E-learning and the first year experience: A framework for best practice. *Educause Australasia*. Retrieved from http://www98.griffith.edu.au/dspace/bitstream/handle/10072/31818/61094_1.pdf?sequence=1.

Ku, H. Y., Akarasriworn, C., Rice, L. A., Glassmeyer, D. M., & Mendoza, B. (2011). Teaching an online graduate mathematics education course for in-service mathematics teachers. *Quarterly Review of Distance Education, 12*(2), 135.

Kubow, P. K., & Fossum, P. R. (2007). *Comparative education: Exploring issues in international context*. Pearson/Merrill/Prentice Hall.

Kujala, J., & Lillrank, P. (2004). Total quality management as a cultural phenomenon. *The Quality Management Journal, 11*(4), 43.

Kyvik, S. (2009). *The dynamics of change in higher education* (Vol. 27). Oslo: Springer.

Lau, R. W., Yen, N. Y., Li, F., & Wah, B. (2014). Recent development in multimedia e-learning technologies. *World Wide Web, 17*(2), 189–198.

Laurillard, D. (2013). *Rethinking university teaching: A conversational framework for the effective use of learning technologies*. Routledge.

Lavrischeva, E., & Ostrovski, A. (2012). General disciplines and tools for e-learning software engineering. *International Conference on Information and Communication Technologies in Education, Research, and Industrial Applications* (pp. 212–229). Berlin Heidelberg: Springer.

Lee, J. (2014). An exploratory study of effective online learning: Assessing satisfaction levels of graduate students of mathematics education associated with human and design factors of an online course. *The International Review of Research in Open and Distributed Learning, 15*(1).

Lee, L.-S. (1998). Four little dragons' technology education. *Eric*. Retrieved from http://eric.ed.gov/?id=ED416382.

Lee, N., & Rozinah, J. (2009). Developing and validating an instrument to assess the e-learning environment at a public higher education institution in Malaysia. *Measuring Business Excellence, 12*(3), 56–75. Retrieved from http://www.emeraldinsight.com/doi/abs/10.1108/13683040810900403.

Lewis, M. J., Davies, R., Jenkins, D., & Tait, M. I. (2005). A review of evaluative studies of computer-based learning in nursing education. *Nurse Education Today, 25*(8), 586–597.

Li, X., & Yan, W. (2012). A comparative study on China-US'APTHS (Academic Proficiency Test for High Schools): Past, present and future. *US-China Education Review*, 534.

Lomas, L. (2004). Embedding quality: The challenges for higher education. *Quality Assurance in Education, 12*(4), 157–165. http://dx.Doi.org/10.1108/09684880410561604.

Lorenzo, G., & Moore, J. (2002). *Five pillars of quality online education*. The Sloan consortium report to the nation, 15–09. Retrieved from http://www.edtechpolicy.orgwww.edtechpolicy.org/ArchivedWebsites/Articles/FivePillarsOnlineEducation.pdf.

Low, C. (2007). Too much e-learning ignores the latest thinking in educational psychology. Retrieved from http://cedma-europe.org/newsletter%20articles/misc/Too%20much%20e-learning%20ignores%20the%20latest%20thinking%20in%20educational%20psychology%20(Jun%2004).pdf.

Madsen, P. (2003). *Debating the future of learning: Exploring online learning in the culture of correctional services organisations*. Flexible learning advisory group. Retrieved from http://www.flexiblelearning.net.au/projects/resources/future-learning-summary.pdf.

Magoulas, G. D., & Chen, S. Y. (Eds.). (2006). *Advances in web-based education: Personalized learning environments*. IGI Global.

Maki, R. H., Maki, W. S., Patterson, M., & Whittaker, P. D. (2000). Evaluation of a Web-based introductory psychology course: I. Learning and satisfaction in on-line versus lecture courses. *Behavior Research Methods, Instruments, & Computers, 32*(2), 230–239.

Marshall, S. J. (2006). *eMM version two process assessment workbook*. Report to the New Zealand Ministry of Education, Wellington: Victoria University of Wellington.

Marshall, S. J. (2012). An analytic framework to support e. learning strategy development. *Campus-Wide Information Systems, 29*(3), 177–188.

Martinez, M. (2010). Adaptive and personalized learning: Supporting individual learning differences. *The Training Place*. Retrieved from http://www.trainingplace.com/source/research/masspersonalization.htm.

Mason, R., & Rennie, F. (2006). *Elearning : The key concepts*. London; New York: Routledge.

Masoumi, D. (2010). *Quality in e-learning in a cultural context: The case of Iran*. Department of Education, Communication and Learning; Institutionen för pedagogik, kommunikation och lärande. Retrieved from https://gupea.ub.gu.se/bitstream/2077/22173/1/gupea_2077_22173_1.pdf.

Masoumi, D., & Lindström, B. (2012). Quality in e-learning: a framework for promoting and assuring quality in virtual institutions. *Journal of Computer Assisted Learning, 28*(1), 27–41.

Maxwell, J. A. (2012). *Qualitative research design: An interactive approach* (Vol. 41): Sage.

Mayadas, F. Miller, G. & Sener, J. (2016) Updated e-learning definitions; definitions of elearning courses and programs version 2.0 April 4, 2015. Retrieved from http://onlinelearningconsortium.org/updated-e-learning-definitions-2/.

McGorry, S. Y. (2003). Measuring quality in online programs. *The Internet and Higher Education, 6*(2), 159–177. doi:10.1016/S1096-7516(03)00022-8.

McLoughlin, C. (1999). Culturally responsive technology use: developing an on-line community of learners. *British Journal of Educational Technology, 30*(3), 231–243. doi:10.1111/1467-8535.00112.

McLoughlin, C., & Visser, T. (2003). Quality and e-learning: are there universal indicators?. In *Sixteenth ODLAA Biennial Forum Conference Proceedings, Sustaining Quality Learning Environments, Canberra* (Vol. 13).

Merriam, S. B. (1988). *Case study research in education: A qualitative approach*. Jossey-Bass.

Merriam, S. B., & Caffarella, R. S. (1999). *Learning in Adulthood: A comprehensive guide education series the Jossey-Bass higher and adult education series*. Wiley.

Miles, M. B., & Huberman, A. M. (1994). *Qualitative data analysis* (2nd ed.). Sage.

Milman, N. B., Posey, L., Pintz, C., Wright, K., & Zhou, P. (2015). Online Master's students perceptions of institutional supports and resources: Initial survey results. *Online Learning Journal, 19*(4).

Miyakawa, H., & Casco, R. A. I. (2007). A comparative study on the curriculum for technology teachers in Japan and Paraguay. *International Journal of Curriculum Development and Practice, 9*(1), 1–14.

Moore, J. C. (2005). *The Sloan Consortium quality framework and the five pillars*. The Sloan Consortium. Retrieved from http://www.mit.jyu.fi/OPE/kurssit/TIES462/Materiaalit/Sloan.pdf.

Moore, J. C., & Fetzner, M. J. (2009). The road to retention: A closer look at institutions that achieve high course completion rates. *Journal of Asynchronous Learning Networks, 13*(3), 3–22.

Moore, M. G., & Kearsley, G. (1996). The theoretical basis for distance education. In *Distance education: A systems view* (pp. 197–212). Elmont, CA: Wadsworth.

Morris, E. (2009). *Cultural dimensions and online learning preferences of Asian students at Oklahoma State University in the United States*. Doctoral dissertation. ProQuest Oklahoma State University-Stillwater.

Mueller, M. M., Palkovic, C. M., & Maynard, C. S. (2007). Errorless learning: Review and practical application for teaching children with pervasive developmental disorders. *Psychology in the Schools, 44*(7), 691–700.

Nayda, R., & Rankin, E. (2009). Information literacy skill development and lifelong learning: Exploring nursing students' and academics' understandings [online]. *Australian Journal of Advanced Nursing, 26*(2), 27–33.

Ndoye, A. (2003). Experiential learning, self-beliefs and adult performance in Senegal. *International journal of Lifelong Education, 22*(4), 353–366. Retrieved from http://www.tandfonline.com/Doi/abs/10.1080/02601370304831.

Newton, D. (2007). *Developing an integrated e-learning culture: a model grounded in the Australian Army experience*. Theses, 62. Retrieved from http://epubs.scu.edu.au/cgi/viewcontent.cgi?article=1064&context=theses.

Nisbet, D. (2004). Measuring the quantity and quality of online discussion group interaction. *Journal of eLiteracy, 1*(2), 122–139.

Nkonge, B., & Gueldenzoph, L. E. (2006). Best practices in online education: Implications for policy and practice. *Business Education Digest*, (15).

Norton, A. (2014). *Mapping Australian higher education*. Grattan Institute report. Retrieved from http://grattan.edu.au/wp-content/uploads/2014/10/816-mapping-higher-education-20141.pdf.

Novak, R. J. (2002). Benchmarking distance education. *New Directions for Higher Education, 2002*(118), 79–92. doi:10.1002/he.57.

Oliver, R. (2001). Developing e-learning environments that support knowledge construction in higher education. In S. Stoney & J. Burn (Eds). *Working for excellence in the e-conomy* (pp. 407–416). Churchlands: Australia, We-B Centre. Retrieved from http://ro.ecu.edu.au/ecuworks/4766/.

Oliver, R. (2005). Quality assurance and e-learning: blue skies and pragmatism. *Research in Learning Technology, 13*(3).

Pappano, L. (2012). Suddenly, millions have started taking U.S. University classes online; Elite schools are working with private providers in offering open courses. *The New York Times, 2* (12), 2012.

Parsazadeh, N., Zainuddin, N. M. M., Ali, R., & Hematian, A. (2013). A review on the success factors of e-learning. In *The Second International Conference on e-Technologies and Networks for Development* (pp. 42–49). The Society of Digital Information and Wireless Communication. Retrieved from http://sdiwc.net/digital-library/a-review-on-the-success-factors-of-elearning.html.

Partridge, H., Ponting, D., & McCay, M. (2011). *Good practice report: Blended Learning*. Australian learning and teaching council . Report ID Code 47566. Retrieved from http://eprints.qut.edu.au/47566/1/47566.pdf.

PASSHE . (2014). *Strategic Plan 2020 (Rising to the Challenge)*. Pennsylvania's State System of Higher Education. Retrieved from http://www.passhe.edu/inside/bog/Documents/Strategic%20Plan%202020%20Rising%20to%20the%20Challenge_10-14.pdf.

Patton, M. Q. (2002). *Qualitative research and evaluation methods*. Sage.

Pemberton, A. & Moallem, M. (2013). The impact of personalized learning on motivation in online learning. In R. McBride & M. Searson (Eds.), *Proceedings of Society for Information Technology & Teacher Education International Conference* (pp. 907–914). Chesapeake, VA: Association for the Advancement of Computing in Education (AACE).Retrieved from http://www.editlib.org/p/48228/.

Phipps, R., & Merisotis, J. (2000). *Quality on the line: Benchmarks for success in internet-based distance education*. The Institute for Higher Education Policy, 1-45. ERIC. Retrieved from http://eric.ed.gov/?id=ED444407.

Putnam, R. T., & Borko, H. (2000). What do new views of knowledge and thinking have to say about research on teacher learning? *Educational Researcher, 29*(1), 4–15.

Raitman, R., Augar, N., & Zhou, W. (2005). Employing wikis for online collaboration in the e-learning environment: Case study. In *Third International Conference on Information Technology and Applications, 2005. ICITA 2005* (Vol. 2, pp. 142–146). IEEE.

Ragin, C. C., & Becker, H. S. (1992). *What is a case?: Exploring the foundations of social inquiry*. Cambridge University Press.

Reeves, T. C. (1992). Effective dimensions of interactive learning systems. *Paper presented at the 2nd Information Technology for Training and Education (ITTE '92) Conference*, St Lucia, Brisbane.

Reeves, T. C. (1994). Evaluating what really matters in computer-based education. *Computer education: New perspectives* (pp. 219–246).

Reeves, T. C., & Harmon, S. W. (1993). Systematic evaluation procedures for interactive multimedia for education and training. *Multimedia computing: Preparing for the 21st century* (p. 472).

Reeves, T. C., & Laffey, J. M. (1999). Design, assessment, and evaluation of a problem-based learning environment in undergraduate engineering. *Higher Education Research & Development, 18*(2), 219–232.

Reeves, T. C., & Reeves, P. M. (1997). Effective dimensions of interactive learning on the World Wide Web. *Web-based instruction* (pp. 59–66).

Reeves, T. D., & Pedulla, J. J. (2011). Predictors of teacher satisfaction with online professional development: evidence from the USA'se-Learning for Educators initiative. *Professional Development in Education, 37*(4), 591–611.

Reich, J. (2014). Harvard measures its MOOCs .*John Harvard's Journal.*. Retrieved from http://harvardmagazine.com/2014/05/harvard-measures-its-moocs.

Rezaei, A. R., & Katz, L. (2002). Using computer assisted instruction to compare the inventive model and the radical constructivist approach to teaching physics. *Journal of Science Education and Technology, 11*(4), 367–380. doi:10.1023/A:1020894002232.

Robert A. Ellis, & Rafael A. Calvo. (2007). Minimum indicators to assure quality of LMS-supported blended learning. *Journal of Educational Technology & Society, 10*(2), 60–70. Retrieved from http://www.jstor.org/stable/jeductechsoci.10.2.60.

Rogers, P. C., Graham, C. R., & Mayes, C. T. (2007). Cultural competence and instructional design: Exploration research into the delivery of online instruction cross-culturally. *Educational Technology Research and Development, 55*(2), 197–217. doi:10.1007/s11423-007-9033-x.

Rosenbaum, P.-E. L. (2013). Educational design: Learners attitudes towards blended learning with online multimedia instruction. *International Journal of Learning and Teaching, 4*(2), 42–61.

Rosenfeld, M., & Rosenfeld, S. (2004). Developing teacher sensitivity to individual learning differences. *Educational Psychology, 24*(4), 465–486. doi:10.1080/0144341042000228852.

Rothkopf, E. Z. (1970). Concept of mathemagenic activities. *Review of Educational Research, 40* (3), 325–336.

Salmon, G. (2005). Flying not flapping: A strategic framework for e-learning and pedagogical innovation in higher education institutions. *ALT-J, Research in Learning Technology, 13*(3), 201–218.

Sangrà, A. (2002). A new learning model for the information and knowledge society: The case of the Universitat Oberta de Catalunya (UOC), Spain. *The International Review of Research in Open and Distance Learning*, *2*(2).

Sangrà, A., Vlachopoulos, D., & Cabrera, N. (2012). Building an inclusive definition of e-learning: An approach to the conceptual framework. *The International Review of Research in Open and Distributed Learning*, *13*(2), 145–159. Retrieved from http://www.irrodl.org/index.php/irrodl/rt/printerFriendly/1161/2146.

Sankey, M. (2014). Benchmarking for technology enhanced learning: taking the next step in the journey. In *Proceedings of the 31st Annual Conference of the Australasian Society for Computers in Learning in Tertiary Education* (pp. 668–672). Macquarie University. Retrieved from http://eprints.usq.edu.au/26449/.

Santos, O. C., & Boticario, J. G. (2015). User-centred design and educational data mining support during the recommendations elicitation process in social online learning environments. *Expert Systems, 32*(2), 293–311.

Scardamalia, M., & Bereiter, C. (2006). Knowledge building: Theory, pedagogy, and technology. In K. Sawyer (Ed.), *Cambridge Handbook of the learning sciences* (pp. 97–118). New York: Cambridge University Press.

Schloss, P. J., & Smith, M. A. (1999*). Conducting research.* Upper Saddle River, NJ: Prentice – Hall, Inc.

Schworm, S., & Gruber, H. (2012). e-Learning in universities: Supporting help-seeking processes by instructional prompts. *British Journal of Educational Technology, 43,* 272–281. doi:10.1111/j.1467-8535.2011.01176.x.

Seidel, N. (2014). Interaction design patterns for design and development of video learning environments. In *Proceedings of the 19th European Conference on Pattern Languages of Programs, 20*. ACM.

Selim, H. M. (2007). Critical success factors for e-learning acceptance: Confirmatory factor models. *Computers & Education, 49*(2), 396–413.

Shailaja, J., & Sridaran, R. (2014). Taxonomy of e-learning challenges and an insight to blended learning. In *2014 International Conference on Intelligent Computing Applications (ICICA)* (pp. 310–314). IEEE. Retrieved from http://www.researchgate.net/profile/R_Sridaran/publication/272151018_Taxonomy_of_E-Learning_Challenges_and_an_Insight_to_Blended_Learning/links/54db8cab0cf23fe133ad4cd9.pdf.

Shopova, T. (2012). E-Learning in higher educational environment.In *International Conference The Future of Education, Italy*.

Skalka, J., Drlik, M., & Svec, P. (2012). E-learning courses quality evaluation framework as part of quality assurance in higher education. In *2012 15th International Conference on Interactive Collaborative Learning (ICL)* (pp. 1–5). IEEE.

Sloman, M. (2002). The Elearning revolution: How technology is driving a new training paradigm. *American Management Association, 222*.

Smerdon, B., Burkham, D., & Lee, V. (1999). Access to constructivist and didactic teaching: Who gets it? Where is it practiced? *The Teachers College Record, 101*(1), 5–34.

Smith, G.G., Passmore, D., Faught, T. (2009). The challenges of online nursing education. *The Internet and Higher Education 12* (2), 98–103. Retrieved from http://dx.doi.org/10.1016/j.iheduc.2009.06.007.

Smith, S., Salaway, G. and Borreson Caruso,J. (2009). *The ECAR Study of Undergraduate Students and Information Technology*. EDUCAUSE Center for Applied Research.

Song, L., Singleton, E. S., Hill, J. R., & Koh, M. H. (2004). Improving online learning: Student perceptions of useful and challenging characteristics. *The Internet and Higher Education, 7*(1), 59–70. doi:10.1016/j.iheduc.2003.11.003.

Spector, J. M., Merrill, M. D., Elen, J ., & Bishop, M. J. (Ed.). (2008). *Handbook of research on educational communications and technology*. Springer.

Spitzberg, B. H., & Hurt, H. T. (1987). The measurement of interpersonal skills in instructional contexts. *Communication Education, 36*(1), 28–45. doi:10.1080/03634528709378639.

Stacey, E. (2002). Learning links online: Establishing constructivist and collaborative learning environments. In *Proceedings of the International Education & Technology Conference: Untangling the web-establishing learning links* (pp. 1–6). ASET.

Stake, R. E. (1995). *The art of case study research*. Sage.

Stemler, S. (2001). An overview of content analysis. *Practical Assessment, Research & Evaluation, 7*(17), 137–146. Retrieved from http://PAREonline.net/getvn.asp?v=7&n=17.

Stockley, Derek. (2003). *How OE-learning definition and explanation (Elearning, Online Training, Online Learning) online learning is revolutionizing K-12 education and benefiting students*. Retrieved from http://www.derekstockley.com.au/elearning-definition.html.

Stodden, R. A., Roberts, K. D., Picklesimer, T., Jackson, D., & Chang, C. (2006). An analysis of assistive technology supports and services offered in postsecondary educational institutions. *Journal of Vocational Rehabilitation, 24*(2), 111–120.

Strategic Plan.(2016). *2016-2020 Strategic Plan*.The University of Sydney. Retrieved from http://sydney.edu.au/dam/intranet/documents/strategy-and-planning/strategic-plan-2016-20.pdf.

Strauss, A. L. (1987). *Qualitative analysis for social scientists*. Cambridge University Press.

Sydney e-learning. (2016). Sydney e-learning. June, 21, 2016 Retrieved from http://sydney.edu.au/elearning/about/index.shtml.

Taber, K. S. (2000). Case studies and generalizability: Grounded theory and research in science education. *International Journal of Science Education, 22*(5), 469-487. Retrieved from http://dx.doi.org/10.1080/095006900289732.

Taylor, J. (1999). Distance education: The fifth generation. In *The 19th ICDE World Conference on Open Learning and Distance Education*, Vienna, Austria.

Taylor, J. (2010). Sustainable higher education learning futures. In *AACE Conference: Global Learn* Asia Pacific, Penang.

Taylor, J. C., & Mackintosh, W. (2011). Creating an open educational resources university and the pedagogy of discovery. *Open Praxis,* (Special Edition, pp. 24–29).

Thomas, R. M. (1993). *International comparative education : Practices, issues, and prospects.* Toronto: International Council for Adult Education.

Thompson, T. (2004). *Results: 2004 survey on access technology in higher education.* Seattle, WA: University of Washington, Access Technologists Higher Education Network.

Thompson, M. M., Braude, E., Canfield, C. D., Halfond, J., & Sengupta, A. (2015). Assessment of KNOWLA: Knowledge Assembly for Learning and Assessment. In *Proceedings of the Second (2015) ACM Conference on Learning@ Scale* (pp. 267–271). ACM.

Thurmond, V. A. (2002). Considering theory in assessing quality of web-based courses. *Nurse Educator, 27*(1), 20–24.

Tobin, K. G. (1993). *The practice of constructivism in science education.* New York: Psychology Press.

Trompenaars, F., & Hampden-Turner, C. (1998). *Riding the waves of culture* . New York: McGraw-Hill. Retrieved from http://ocan.yasar.edu.tr/wp-content/uploads/2013/09/Riding-the-waves_Part-1.pdf.

UKIBC, (2015) . *Meeting India's education challenges through E-Learning.* The Uk India Business Council Supports Uk Businesses Report . Retrieved from http://www.Klik2learn.Com/Wp-Content/Uploads/2015/02/Ukibc_E-Learning-In-India_Report.Pdf.

United States vs Australia. (2016). America vs Australia - cultural differences—Convict Creations. Retrieved from http://www.convictcreations.com/culture/yankaussie.htm.

Usun, S. (2004). Factors affecting the application of information and communication technologies (ICT) in distance education. *Online Submission, 5*(1). http://eric.ed.gov/?id=ED49444.

Van der Vyver, A. G., Williams, B., & Marais, M. A. (2015). Using social media as a managerial platform for an educational development project: Cofimvaba. *International Journal of Information and Education Technology, 5*(12), 910. doi:10.7763/IJIET.2015.V5.636.

VET, G. C. o. A. (2012). National VET E-learning Strategy. In 2012 (Ed.), *Department of industry, innovation, science, research and tertiary education*: Commonwealth of Australia.

Volery, T., & Lord, D. (2000). Critical success factors in online education. *International Journal of Educational Management, 14*(5), 216–223.

Wahlstedt, A., Pekkola, S., & Niemelä, M. (2008). From e-learning space to e-learning place. *British Journal of Educational Technology, 39*(6), 1020–1030.

Walker, C. O., Greene, B. A., & Mansell, R. A. (2006). Identification with academics, intrinsic/extrinsic motivation, and self-efficacy as predictors of cognitive engagement. *Learning & Individual Differences, 16*(1), 1–12. doi:10.1016/j.lindif.2005.06.004.

Walker, R. (1980). The conduct of educational case studies: Ethics, theory and procedures. *Rethinking Educational Research,* 30–63.

Wang, Y. S., (2003). Assessment of learner satisfaction with asynchronous electronic learning systems. *Information and Management, 41*(1), 75–86.

Washburn, E. (2012). *Korean student's online learning preferences and issues: Cultural sensitivity for western course designers.* Doctoral dissertation. Oklahoma State University.

Watson, J. C., & Simpson, L. (2010). *Counseling research: quantitative, qualitative, and mixed methods.* Boston: Pearson.

Weeks, K. W., Clochesy, J. M., Hutton, B. M., & Moseley, L. (2013). Safety in numbers 4: The relationship between exposure to authentic and didactic environments and Nursing Students' learning of medication dosage calculation problem solving knowledge and skills. *Nurse Education in Practice.*

Weld, D. S., Adar, E., Chilton, L., Hoffmann, R., & Horvitz, E. (2012). Personalized Online Education—A Crowdsourcing Challenge. Paper presented at *the Workshops at the Twenty-Sixth AAAI Conference on Artificial Intelligence.*

Weng, C., Tsai, C. C., & Weng, A. (2015). Social support as a neglected e-learning motivator affecting trainee's decisions of continuous intentions of usage. *Australasian Journal of Educational Technology*. Retrieved from http://ajet.org.au/index.php/AJET/article/view/1311.

Wiemann, J. M. (1977). A description of competent and incompetent communication behavior. Retrieved from http://eric.ed.gov/?id=ED147902.

Wild, M., & Henderson, L. (1997). Contextualizing learning in the World Wide Web: Accounting for the impact of culture. *Education and Information Technologies, 2*(3), 179–192. doi:10.1023/A:1018661213335.

Wilkinson, A., Forbes, A., Bloomfield, J., & Gee, C. F. (2004). An exploration of four web-based open and flexible learning modules in post-registration nurse education. *International Journal of Nursing Studies, 41*(4), 411–424.

Wills, S. (2012). Role-based e-learning for university students: A comparison of Australian, American, British and Singapore designs. *Journal of Comparative Eduation.* Retrieved from http://ro.uow.edu.au/cgi/viewcontent.cgi?article=1263&context=asdpapers.

Wirth, M. A. (2005). EFMD Certification of e-Learning (CEL).

Witrant, E., & Popescu, E. (2011). A problem-based learning approach for green control and IT in a master program. *International Conference on Web-Based Learning* (pp. 273–282). Berlin Heidelberg: Springer.

Wlodkowski, R. (1999). *Enhancing adult motivation to learn: A comprehensive guide for teaching all adults*. San Francisco, CA: Jossey-Bass Publishers.

Wolhuter, C. (2008). Review of the review: Constructing the identity of comparative education. *Research in Comparative and International Education, 3*(4), 323–344.

Woo, M. A., & Kimmick, J. V. (2000). Comparison of Internet versus lecture instructional methods for teaching nursing research. *Journal of Professional Nursing, 16*(3), 132–139.

Yazdani, J., & Hassani, M. (2011). A study on the level of conformity between the objectives of curriculum guidelines at the elementary and middle school and official objectives of the supreme council of education. *Journal of curriculum studies, 5*(20), 58–79.

Yeung, D. (2002). Toward an effective quality assurance model of web-based learning: The perspective of academic staff. *Online Journal of Distance Learning Administration*, *5*(2). Retrieved from http://www.westga.edu/~distance/ojdla/summer52/yeung52.htm.

Yin, R. K. (2014). *Case study research: Design and methods* (5th ed.). Los Angeles, CA: Sage.

Young, L. E., & Paterson, B. L. (2007). *Teaching nursing: Developing a student-centered learning environment:* Lippincott Williams & Wilkins. United States of America.

Zhao, F. (2003). Enhancing the quality of online higher education through measurement. *Quality Assurance in Education, 11*(4), 214–221.

Zhou, Y. (2012). Towards Capability Maturity Model of e-Learning Process. *Intelligent Information Management*, *4*(4).

Kolibri 360 GmbH
Ferdinand-Jühlke-Straße 7
99095 Erfurt, Deutschland
produktsicherheit@kolibri360.de